REVOLUTION

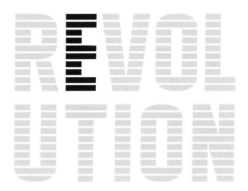

REVOLUTION

TAME TECHNOLOGY_
GET YOUR LIFE BACK_

KEVIN DUNCAN_

HODDER EDUCATION
AN HACHETTE UK COMPANY

For UK order enquiries: please contact Bookpoint Ltd, 130 Milton Park, Abingdon, Oxon OX14 4SB. Telephone: +44 (0) 1235 827720. Fax: +44 (0) 1235 400454. Lines are open 09.00–17.00, Monday to Saturday, with a 24-hour message answering service. Details about our titles and how to order are available at www.hoddereducation.co.uk

British Library Cataloguing in Publication Data: a catalogue record for this title is available from the British Library.

First published in UK 2011 by Hodder Education, part of Hachette UK, 338 Euston Road, London NW1 3BH.

This edition published 2011.

Copyright © 2011 Kevin Duncan

In UK: All rights reserved. Apart from any permitted use under UK copyright law, no part of this publication may be reproduced or transmitted in any form or by any means, electronic or mechanical, including photocopy, recording, or any information, storage and retrieval system, without permission in writing from the publisher or under licence from the Copyright Licensing Agency Limited. Further details of such licences (for reprographic reproduction) may be obtained from the Copyright Licensing Agency Limited, of Saffron House, 6–10 Kirby Street, London EC1N 8TS.

Illustrations by Redmoor Design.

Front cover: © Digital Vision/Getty Images

Typeset by MPS Limited, a Macmillan Company.

Printed in India.

The publisher has used its best endeavours to ensure that the URLs for external websites referred to in this book are correct and active at the time of going to press. However, the publisher and the author have no responsibility for the websites and can make no guarantee that a site will remain live or that the content will remain relevant, decent or appropriate.

Hachette UK's policy is to use papers that are natural, renewable and recyclable products and made from wood grown in sustainable forests. The logging and manufacturing processes are expected to conform to the environmental regulations of the country of origin.

Impression number   10 9 8 7 6 5 4 3 2 1

Year   2015 2014 2013 2012 2011

DEDICATION

As always, my writing is dedicated to my girls: Sarah, Rosanna and Shaunagh.

Big thanks to Dave Hart for the format suggestion.

Also by the author:

Run Your Own Business [Teach Yourself]

Small Business Survival [Teach Yourself]

Business Greatest Hits

Marketing Greatest Hits

So What?

Start

Tick Achieve

For apps, podcasts, support, training products and comment:

expertadviceonline.com

kevinduncan@expertadvice.co.uk

greatesthitsblog.com

00: CONTENTS

FOREWORD	x
INTRODUCTION	xii

PART I: EVOLUTION TO REVOLUTION

1 THE ROAD TO EDICTION — 5
  _ The evolution of ediction
  _ Repetitive strain injury
  _ Technology milestones
  _ The rise of Bleisure time
  _ Enough: have we over-consumed?
  _ Affluenza and ediction
  _ Phobology and self-esteem
  _ Are you in love with a machine?

2 DO YOU NEED A REVOLUTION? — 35
  _ 'I can't talk now, I'm on the phone…'
  _ Yr msgs r kllg me: the tyranny of text
  _ Thousands of emails but no conversations?
  _ Distraction or action?
  _ Everything louder than everything else
  _ It's all the (techno) rage
  _ Hiding behind technology
  _ Can't get no satisfaction (even in topless meetings)
  _ Need a revolution? Take the test

PART II: REVOLUTION RESOLUTIONS

3 THINKING — 69
- Thinking, talking, communicating, doing and being
- Mind games
- Thinking is free, so do it more often
- Small thoughts are as valuable as big ones
- Thinking or just rearranging your prejudices?
- How to out-think yourself
- Your locus of control
- Turn on or turn off?
- Taming Technology Tips for Thinking

4 TALKING — 97
- The art of brachylogy
- Thinking and talking straight
- Spotting waffle
- The spontaneous word dump
- How to listen properly
- How to say 'no' politely
- Time for a new language
- Taming Technology Tips for Talking

5 COMMUNICATING — 127
- The death of human interaction?
- Room to hide
- Don't assume you can handle it all at once
- The right medium for the right message
- How to write concisely and clearly
- Technology morals
- A new communication etiquette
- Taming Technology Tips for Communicating

6 DOING — 157
- Human beings and human doings
- Confronting the devil
- Progress not perfection

_ Action not activity
_ Multitasking and Rapid Sequential Tasking
_ Anti lists
_ When doing nothing is best
_ Taming Technology Tips for Doing

7 BEING 185
_ Defined by your character or your technology?
_ In pursuit of eudemonia
_ Getting your attitude right
_ Humility + honesty + humour = happiness
_ Duration, variation and vacation
_ Taming Technology Tips for Being

RESOURCES 207
_ References
_ Appendix: Cliché and Jargon Red Alert List

INDEX 215

FOREWORD

When I was at school, we lived through a calculator arms race: bigger and bigger, more and more functions, bigger and bigger memory. Many of them are now museum pieces and the size and weight of a small laptop but at the time they were the bleeding edge of technology.

Would the outsourcing of calculation make us stupid, teachers wondered? How could they ensure it was the student and not the machine that did the work? How could they teach us to be smart users rather than dumb slaves of the machines?

In this book, Kevin Duncan takes apart the difficulties that many of us have in dealing with the ubiquitous technology of our age, and gives us simple tips and advice on how to learn to live more comfortably with all the gadgetry that we have come to depend on.

How not to get 'muttered' and 'facetubed' into stupidity.

Brilliantly written, and a must-read.

Mark Earls, author of *Herd*

'KEVIN DUNCAN REALLY NAILS THE PROBLEM MOST OF US FACE IN DEALING WITH THE TECHNOLOGY THAT IS SHAPING OUR WORLD. AND SHOWS YOU SIMPLE AND USEFUL WAYS TO GET TO GRIPS WITH IT. FORGET *WIRED*, BUY THIS.'

Mark Earls

INTRODUCTION

In a dictionary as recently as 2003, technology is described as the application of practical sciences to industry or commerce, or the total knowledge of human skills available to any human society. These days it would include a definition based on devices that save time, enable new ways of communicating, and connect communities all over the world.

Things are changing, and they are changing fast.

Technology can be a force for good – that is not in doubt. But it can also distort perspectives of what life should be all about, and that's what this book tries to address. The internet and mobile phones have become universally available in the last 10 to 15 years, which means that the world now contains millions of people who have never experienced anything else.

In training, anyone under the age of 30 is prone to ask what people did all day without these two developments. The answer, of course, is that the technology of the day was used to make arrangements and get things done, and that meant using the landline, making arrangements in advance, and sticking to them. And we had more time.

This book is not about hankering after the past. It's about dealing with the present. I meet too many people who are not coping with what technology is throwing at them. Too many calls, too many emails. No time to think or relax. In some cases, it has not even occurred to them that they can turn their machines off.

The equation between man and machine has to be right. If you can do more, and faster, with technology, then that should not become a reason to do even more. It should create a moment in which other activities can be pursued. As the saying goes, the moment you achieve the impossible, your boss will simply add it to your regular list of duties. No, when the job is done, it's done, and it should be a matter of human choice what we do next.

It's your life, so don't let your machines run it.

Kevin Duncan, Westminster, 2011

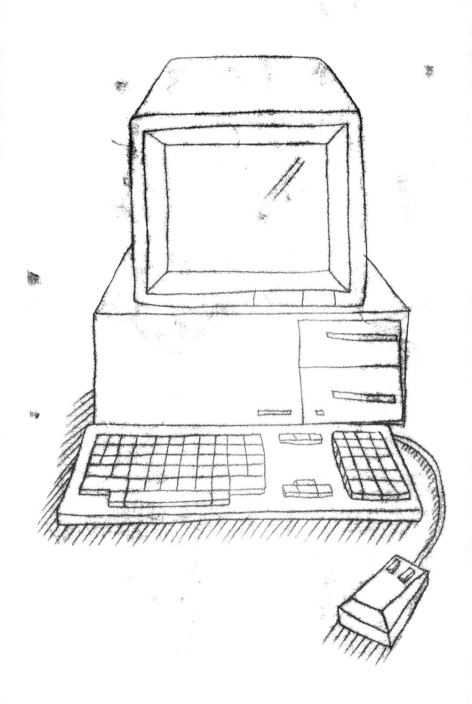

PART I: EVOLUTION TO REVOLUTION

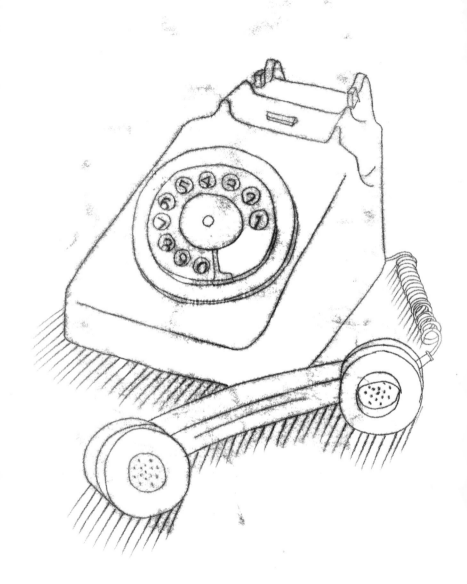

01: THE ROAD TO EDICTION

It's time for us **humans** to fight back

THE EVOLUTION OF EDICTION

> I SIGNED ON, ALTHOUGH I NEVER REALISED IT, FOR 100,000 HOURS OF WORK DURING MY LIFETIME.... MY TEENAGE SON OR DAUGHTER, A GENERATION LATER, CAN EXPECT THEIR JOBS TO ADD UP, ON AVERAGE, TO 50,000 HOURS.
>
> Charles Handy, 1989

Glance back at any business book in the seventies or eighties, and you will find extremely bright people like Charles Handy analysing the manner in which technology would revolutionize our lives. The predictions were all the same: machines would do wonderful things to save us the work, and as a result we would have much more free time – this example suggests 50 per cent more. Well here we are in the future, and it hasn't quite worked out like that. In fact, we work longer hours than ever. So how did it come to this?

You have probably seen the cartoon sequence that shows man rising from an ape to *Homo erectus*, and then regressing down to a slouch until he finally comes to rest on a sofa or in front of a computer. Starting with fire, tools and the wheel, the schematic suggests that technology made man the most powerful species on earth, and then weighed him back down again. It's probably a bit much to call this the ascent and descent of man, but the time has come to review our relationship with technology. The industrial revolution was one thing, the internet revolution was another, and since the invention of the World Wide Web, mobile phones, personal computers, and every other device you can think of, man has been confronted with a lot more than he originally bargained for.

Don't get me wrong: technology is not a bad thing in its own right, but individuals must be able to manage it properly if they are to lead decent, balanced lives. EDICTION is my word for being addicted to something electronic. If you are EDICTED to one device or several, then chances are your levels of stress will be higher than necessary. Not being able to cope with all that the world throws at us technologically is no disgrace, but there seems to be very little help at hand to suggest what people can do to redress the balance, and that's where this book comes in.

After a short run-up to establish how on earth we arrived here, we will try to find out whether you need to start your own revolution – a complete overhaul of how you approach the technology in your life. The idea is relevant to young and old. Older people can learn new skills fast, and a huge number of silver surfers are already proving that. But equally we now have a whole young generation who have never experienced life *without* computers and mobile phones, and many are missing out on life skills as a result. They may wonder what people did with their time before all this technology existed. Going nostalgically dewy-eyed about playing in the street and building tree houses is not the point here – we need the ability to experience both worlds in a complementary and balanced way.

Technology can be a wonderful thing, and has transformed our lives. But it can also be a curse when it overwhelms us. If your phone, computer or other devices are beginning to rule your life, then you may be EDICTED. It doesn't have to be like this. It's time for us humans to fight back. But first, let's see how we got here.

IF FACEBOOK WERE A COUNTRY IT WOULD BE THE THIRD LARGEST IN THE WORLD

REPETITIVE STRAIN INJURY

> THE BIG PICTURE IS MORE LIKELY TO PARALYSE THAN TO INSPIRE.
>
> Matthew Parris

The world arguably has too much information, and many individuals can't cope. In 2007 the digital universe equalled 281 billion gigabytes of data, or about 45 gigabytes for every person on earth. If that doesn't mean much to you, then you might be interested to learn that it was the first time that the overall size of digital content went beyond the total storage capacity. It is predicted that in 2011 only half of the digital world will be stored — the rest will be in transit.

Eye-watering statistics like this keep coming. An exabyte is 1.074 billion gigabytes. Two exabytes equal the total amount of information generated in 1999. The internet currently handles one exabyte of data *every hour.* Inevitably, this statistic will be out of date by the time you read this. We are almost running out of the right language to describe what is happening, so let's strip it back to basics. If individuals can't cope with too much inbound information, then they need to pause and make sure the technology is working for them, and not the other way round. Less is more. Understanding is everything.

The fact that we are generating so much data is both good and bad. If sought and used judiciously, it can be a rich source of insight, and can lead to what Clay Shirky calls a Cognitive Surplus. For the first time ever young people are watching less television than their elders, and are instead using more of their free time for active participation in social interaction online. But many people are so swamped with information that they simply don't know what to do with it, and that can lead to paralysis. In 2009, the world generated more data than in its entire existence beforehand. That truly is a lot to take in. When people talk of information overload, they have a point. There is a limit, and we may well have reached it. Take the number of people in social networks for example. If Facebook were a country it would be the third largest in the world behind China and India, with 500 million active users. That's an enormous community.

The world population is growing by about 1 billion every 13 years, and everybody from Masai tribesmen to Mumbai students wants a phone or a computer, so we have effectively passed the point of no return when it comes to the exponential spread of technology. None of this is going to go away. If anything, it is more likely to increase in speed and intensity. That's the big backdrop. Now let's look at the very recent past.

More mobiles than people

TECHNOLOGY MILESTONES

> EXPERTS BELIEVE THE IPAD WILL REVOLUTIONISE THE WAY WE PROCRASTINATE.
>
> David Letterman

Humans have always had an insatiable need to invent things. Once we had the basic designs for transport, there was no stopping us: boats, airplanes, cars and space rockets. By 1990, you might have had a computer at work, and, heaven forbid, one at home. Posh executives were having phones put in their cars, which involved ripping out half of the interior. By the mid-nineties, mobile phones were becoming common. Fast forward to today, and there are more mobiles in the UK than there are people, and at the last count 4.8 billion people in the world have got one – that's two-thirds of the population. It seems we all love a gadget, whether for work or play. In fact, the first ten years of the current millennium are often referred to as the Decade of the Gadget, and the top ten most popular, in chronological order, were:

▶ **USB stick (2000)**
IBM launched the DiskOnKey, thereby condemning the floppy disk to the history books, and the bin. It soon became the quickest way to transfer files between computers.

▶ **Apple iPod (2001)**
This machine has changed utterly the way in which we consume music. Originally only capable of holding

1,000 songs, it has now become a multi-media receptacle.

▶ Sky+ (2001)

No need for tapes, or careful planning and programming of recorders. This easy-to-use box with a large hard drive allowed viewers to record hundreds of hours of programmes at the touch of a button.

▶ BlackBerry (2002)

Loathed and revered in equal measure, this mobile email device meant life would never be the same for many. Fans love the 'always on' functionality – families and partners less so.

▶ TomTom Go (2004)

Goodbye maps, hello Satellite Navigation. Ideal for those with no sense of direction, until they take it too literally and end up driving into a lake.

▶ Slingbox (2005)

A television streaming device that allows you to shift (sling) shows from your TV or video recorder to a computer anywhere in the world.

▶ Nintendo Wii (2006)

This family-orientated games console arguably introduced a whole new generation to video games. 56 million sold and rising.

▶ Flip (2007)

A pocket-sized camcorder that can capture about an hour of footage. The built-in USB stick flips out, allowing editing and uploading to the web. Much loved by YouTube users, now anyone can be an on-the-spot journalist.

▶ Asus Eee (2007)

Not the world's most well-known brand name, but this ultra-light laptop kickstarted the netbook genre – a new category of basic computers for sending emails and using the internet on the go, usually for less than £200.

▶ iPhone (2007)

The mobile phone was now totally redefined, and responsible for more than a billion apps being downloaded since the launch of the App store in 2009.

This extraordinary list almost needs no further comment. Suffice to say that these gadgets, and scores of others, have completely changed the way in which humans operate, and if you were born in the last 10 or 20 years, all this is the norm. Many trainees have never heard of a fax, and a recent 22-year-old recruit to Apple had never used email because his life was run purely by mobile phone and social media. Indeed, the role of machines is so profound that they have even blurred the distinction between what is leisure and what is business.

We are all born to play and create

THE RISE OF BLEISURE TIME

> IT'S NOT THE PRETTIEST WORD, BUT YOU'D BETTER GET USED TO IT BECAUSE THIS BLURRING BETWEEN OUR WORKING LIVES AND DOWNTIME, OR BUSINESS AND LEISURE, IS HERE TO STAY.
>
> Tony Turnbull

The rise of BLEISURE can be viewed in two ways. If we choose to let it be, our world is 'always on'. That's good if you want to play music and watch films, and bad if you can never mentally leave the office. If you turn your iPhone on to download some music, will you see the new stream of office emails? The tools of work and play have effectively become one, and the traditional nine-to-five has ceased to exist. People take business calls in the evening and visit eBay from their desks, if the company system lets them.

Separating life and work has become a lost skill, but maybe this should not come as a surprise, because we are all born to play and create, according to Pat Kane, author of *The Play Ethic*. He suggests that politicians arguing consistently for a work ethic are missing the point. We are essentially designed to play. We all think we know what play is (what we do as children, outside work, and for no other reason than pleasure), but understanding the real meaning of it would revolutionize and liberate our daily lives. Play offers learning, progress, imagination, a sense of self, identity and contest. It is also the fermenting ground for exploring alternatives – the very essence of creativity.

Huge numbers of companies have caught on to this idea and now make their money out of play elements. This is good news if a new generation of workers can get paid for doing something they enjoy, and the more work is like play, the more likely this is to be the case. But technology needs careful handling where the work ethic slams into the play ethic. Millions of working hours are lost in companies around the world as employees muck about on websites, chat rooms, social media and more. Companies try to stop it, but only partially succeed.

The idea of work/life balance has become confusing. The difference between work and play in a technological world has become blurred. You would have thought that those who spend all day at work on a computer would want to do something else in their spare time, but often they just do more of it when they get home. Using technology can be addictive, and doing so is a form of consumption in its own right.

What was I looking for?

ENOUGH: HAVE WE OVER-CONSUMED?

AM I BEING UP-GADGETED? MASSES OF PLANETARY RESOURCES GET PILED INTO ADDING FEATURES TO MAKE OUR LIVES EASIER. BUT DO THEY REALLY ADD ANYTHING? IF THEY DON'T, THEN WE ARE JUST SPENDING MONEY ON LANDFILL.

John Naish

John Naish, author of *Enough*, points out that our basic survival strategy makes us chase more of everything: status, food, information and possessions. Now, thanks to a mixture of technology and money, we have suddenly got more of everything than we can ever use. As a result, we urgently need to develop a sense of 'enough', and an ability to enjoy what we have, rather than fixating on 'more'. This frequently applies to the purchase of clothes, for example, where in truth we only need a certain amount but often we just keep on filling the wardrobe.

Scientists have shown that purchasing items gives us a dopamine rush but it wears off almost immediately. Thousands of women in particular routinely return everything on a Monday that they bought on Saturday. Retailers call them shoe-limics. **WILF**ING is a pointless form of shopping: *What Was I Looking For?* In this case, the activity has become the point, not the purchase. Technology addicts can suffer from the same phenomenon, buying more and more gadgets whether they are 'needed' or not. 'ALL THE GEAR BUT NO IDEA' is how one sports equipment shop owner describes his fair-weather customers, and the phrase could equally apply to anyone with too many technological devices. These people are most likely EDICTED.

When it comes to data and information, many of us are suffering from INFOBESITY. Too much information causes stress and confusion and makes people do irrational things. For example, our 24-hour news media suffers from an 'ELVIS STILL DEAD' syndrome that distorts our view of the world to the point that many of us have effectively forgotten what true news is. We keep on consuming it until we become bloated with too much information.

Developing a sense of enough is also necessary at work. As we saw at the beginning of the book, we were expected to be working less by now, thanks to the power of technology, but actually we are working more. This is partly because of the blurriness of Bleisure time, but also because some of us have developed a warped attitude to work. Presenteeism is a phenomenon whereby people spend hours at their desks not achieving anything because they are too tired, stressed, under-stimulated, distracted or depressed to be productive. Workaholics Anonymous is a movement based on Alcoholics Anonymous principles. Ironically, when they tried the scheme in New York, only two people turned up because the rest were too busy. In another extreme paradox, earning more simply increases discontentment. Many people believe (or convince themselves) that their overwork habits are driven by irresistible external forces. They then frown on normal timekeepers and make their lives a misery.

Has all this consumption made people any happier? No it hasn't, but amazingly people always think things will be better in the future. Present quality of life is deemed to be 6.9 out of 10 (and guessed at 8.2 in 5 years time). But when the time comes, it's still 6.9. In many instances, this over-consumption is a case of too much too soon, and in some cases, too much, full stop. Customer demands are being met fast, and followed immediately by the creation of, and satiating of, even more technological demands. Some gadgets appear to fill needs that people previously never even knew they had.

Are you constantly up-gadgeting? 60 per cent of adults only use less than half of the functions on their devices. If any of this rings a bell with regard to you and your approach to technology, then you may be EDICTED.

AFFLUENZA AND EDICTION

> TEN YEARS AGO, WE'D HAPPILY QUEUE IN A BANK TO GET £50. NOW, IF YOU'RE STANDING AT A CASH MACHINE AND ONE PERSON IS PRESSING THE RECEIPT BUTTON, YOU ACTUALLY WANT TO KILL THEM.
>
> Simon Cowell

It is now beyond doubt that we buy more technology than we need. We want a lot and we want it fast. AFFLUENZA is defined by Oliver James as a contagious middle-class virus causing depression, addiction and ennui. This is an epidemic sweeping the world, and in order to counteract it and ensure our mental health, we should pursue our needs rather than our wants – the majority of which are unsustainable.

Most middle-class people have too much of everything, but it hasn't made them any happier. People need to reject much of the status quo in order to be a satisfied, unstressed individual. We have become more miserable and distressed since the seventies, thanks to successive governments pushing the cause of personal capitalism. While there has been a massive increase in the wealth of the wealthy, there has been no rise in average wages. We need to recapture a sense of self-worth and personal wellbeing if we are to overcome these negative feelings. Erich Fromm's theory of American consumerism said that the choice in the fifties was 'TO HAVE OR TO BE', and that people have become Marketing Characters 'based on experiencing oneself as a commodity'. Arguably, nothing much has changed.

Technology can help with much of this if used wisely. Equally, it is worth remembering that humans only really have four basic emotional needs: to feel safe and secure, to feel competent, to feel connected to others, and to feel autonomously and authentically engaged in work and play. Earning more creates a desire for more technology, and self-doubt correlates with materialism. Interestingly, everyone feels that 'enough' income is always 10 per cent more than they have, and the range of goods regarded as 'essential' in a household has increased dramatically, as has the speed with which they can be obtained – in an on-demand world, everybody wants everything now. Technology feeds this need and everybody is getting impatient. Time is being squeezed in a concertina. The faster it comes, the faster you want it next time, but as Gandhi pointed out: '*There is more to life than increasing its speed.*'

Too much choice
Too much choice
Too much choice
Too much choice
Too much choice
Too much choice
Too much choice
Too much choice
Too much choice
Too much choice

PHOBOLOGY AND SELF-ESTEEM

> TECHNOLOGY IS A QUEER THING: IT BRINGS YOU GREAT GIFTS WITH ONE HAND, AND STABS YOU IN THE BACK WITH THE OTHER.
>
> C.P. Snow

Phobology is the study of phobias, and we have a lot of them. The top five phobias in the UK, in reverse order, are driving (1%), erythrophobia – that's blushing (1.2%), emetophobia – vomit (2.6%), agoraphobia (9.9%), and at number one is social phobia on 17.2%. Technology has played a role in this, both good and bad. On the plus side, it has opened up networks for shy people, allowed internet dating, helped those with learning difficulties, provided instant messaging capability for those in therapy who don't want to talk on the phone, and much more. On the minus side, it can allow people to hide behind it, reduce face-to-face interaction and build barriers in relationships. As Erik Qualman claims in his book *Socialnomics*, 'THE NEXT GENERATION CAN'T SPEAK'. If you would rather email your wife than talk to her, then things may be going wrong.

The launch of endless new products can lead to paranoia and reduced self-esteem. More and more choice might mean 50 brands of cooking oil, 200 brands of beer, 500 TV channels, and tens of thousands of websites. This all sounds rather wonderful at first glance, but the American sociologist Barry Schwarz has studied product proliferation and believes that, after a certain point, too much choice overloads our brains. 'INCREASED CHOICE MAY ACTUALLY CONTRIBUTE TO THE RECENT EPIDEMIC OF CLINICAL DEPRESSION AFFECTING MUCH OF THE WESTERN WORLD', he says in *The Paradox of Choice*. He was rather dismayed to find 16 varieties of instant mashed potatoes in his local supermarket. Science writer James Gleick reached the same conclusion, lamenting that the more telephone lines you have, the more you need. The complications beget choice; the choice inspires technology; the technologies create complication. It's a never-ending cycle.

Keeping up with the Joneses has always been a messy and soul-destroying business, and proliferation of choice makes it worse. Consider the role of technology in fuelling this phenomenon. A household in the fifties might aspire to a radio, a fridge, and possibly a food mixer. The average kitchen now has every conceivable feature, and may even include a fridge with an integrated digital shopping list. One television set per house is now inconceivable, and there may well be one in your car or in your hand. The screen in your living room could well be an all-in entertainment system. Many people have several phones, and more than one version of what is effectively the same device.

Convergence brings convenience and pleasure on the one hand, and heartbreak and stress on the other. How convenient it is to have your phone, email, contacts, messages, important data, diary and hundreds of other vital items on your hand-held device – until you lose it. 'I COULDN'T LIVE WITHOUT MY PHONE', is the mantra of the modern generation. Few have landlines, alarm clocks, watches, diaries or any other old-fashioned paraphernalia. Their whole life is concentrated on one omnipotent device – until it gets stolen, falls down the toilet, or just stops working. This is not necessarily a tale of woe. It just needs putting in perspective. Self-esteem and inner confidence should come before total dependence on any piece of technology. No one should depend on any single thing that much, be it human or machine. Which begs the question: do you?

like a tidal wave

ARE YOU IN LOVE WITH A MACHINE?

THE ODDS ARE GOOD BUT THE GOODS ARE ODD.
Online dating customer

There's nothing wrong with online dating, and it has certainly generated high levels of satisfaction for many looking for a partner. As with many things in life, it works well if handled sensitively, notwithstanding the witty observation that the odds are good but the goods are odd. On reflection, that could just as easily be the case when dating anyone, regardless of where you met. Falling in love when you have met someone is great, but is it okay to be in love with a machine? Do you know anyone who is? Are you yourself? These questions aren't as daft as they may sound. Humans, and particularly men, have always loved machines. There are devoted enthusiasts for trains, planes, cars, boats, motorbikes and hundreds of others. It's just that technological advances have made many of these machines more appealing to both sexes, to young and old, and have made them pervade every walk of life. Many relationships have been lost to those with obsessive hobbies that drive their partners mad and make them feel excluded, and modern devices have exacerbated the problem.

Computer games, mobile phones, personal organizers, laptops, email, or even just the ability to work at home have ripped apart many a partnership. So now when someone stares admiringly at their iPad, is it love? It might well be, and if it is, it could be a dangerous moment for the admirer. And there are probably a lot of them. To put this in perspective, it took Apple over six years to sell a million of its Apple II computers after launching in 1977. The iMac achieved that milestone in just over a year in 1998, as did the iPod in 2001. The iPhone sold a million in 74 days in 2007 and the iPad did it in 28 days. Over 5 million have been shifted so far, although there is much debate about their intrinsic value and purpose. James Lileks drily noted that 'THE IPAD IS THE BARACK OBAMA OF TECHNOLOGY. IT'S WHATEVER YOU WANT IT TO BE, UNTIL YOU ACTUALLY GET IT.'

There is nothing wrong with these sales figures, nor the machines. In fact, they are brilliantly clever. I own plenty of them myself. The critical question is what they mean to the individual. If they mean connection, freedom of expression, ability to interact with others as never before, and a host of other benefits, then clearly that's a good thing. All that benefit, however, relies on the user keeping a balanced approach to how much time they spend interacting with the technology, whether that represents a sensible ratio in relation to pursuing other activities, and whether those around them – friends, partners and family – find that level of interaction healthy and acceptable.

All of which leaves the modern human with much to ponder. Technology has always been creeping up on us, but now it has come upon us like a tidal wave. The new digital natives embrace it with open arms and know no other way of living. Those who are a little older may find much of it daunting. There is no system with which to judge what a sensible level of involvement with one's technology might be, so it falls to the individual to take a sensible view.

You can start by asking yourself whether you are EDICTED. If you are not sure, then the symptoms suggested in the next chapter might help you decide.

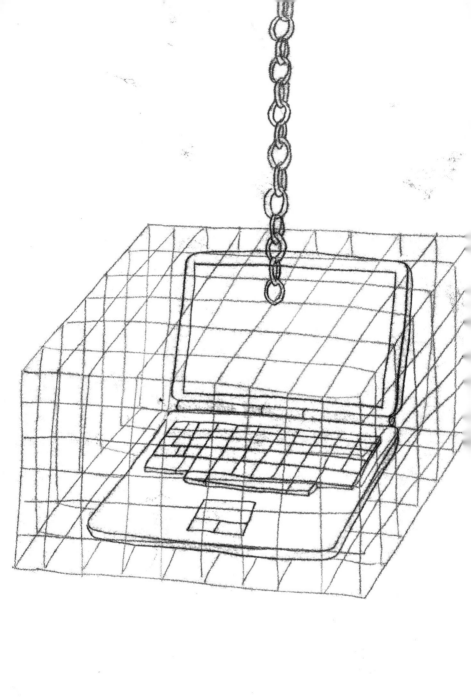

02: DO YOU NEED A REVOLUTION?

So we know that technology can cause a lot of trouble if it remains untamed. If a person is addicted to technology then they are EDICTED. Is this you? Are you controlling your technology or is it the other way round? You can take the test at the end of this chapter to find out how bad things are, and decide whether you need to start a revolution in your life. But first let's look at some of the main culprits.

NOT A CONVERSATION AT ALL —

"I CAN'T TALK NOW, I'M ON THE PHONE..."

Do you ever interrupt a conversation with your partner or children to take a call?

You walk into the reception area of a hotel or business, and it often happens. You are talking to the person, the phone rings, and they break off your conversation mid-sentence to answer it. It's rude, and it demonstrates the power of interruption by technology. Why would anyone believe that the inbound intervention would be a better option than the real conversation they are currently engaged in?

In the modern world, we can all be guilty. We used to have two-way conversations when one person spoke to another, and their concentration was total. Now we often have three-way conversations in which one party is probably fiddling with a device, and in some cases both might be, so it is really a four-way conversation, or arguably not a conversation at all. As *The Times* reported of one frustrated partner at their wedding anniversary lunch: 'IF YOU TALK TO ME, I EXPECT EYE CONTACT. MEANWHILE YOU ARE TYPING SOME MEANINGLESS OBSERVATION INTO THE ETHER.' It has been said that mobiles and the internet have opened up the world on the one hand, only to shrink our horizon to two inches on the other – the size of the screen on our devices.

Research shows that many people now regularly keep their devices by their beds at night. Relationship counsellors talk of marriages breaking up through lack of consideration. Many partners feel they are being excluded if their partner is spending a lot of time using phones for socializing, playing games or working. Psychologists have observed that there is something quite compelling about contemporary gadgetry. Modern gadgets activate a part of the brain that wants to be absolutely absorbed. This creates a strange altered state in which the user is with their partner physically but not available to them mentally.

New symptoms have been created. OBSESSIVE MOBILE DISORDER has been joined by CONTINUOUS PARTIAL ATTENTION (or CPA), in which victims come to believe that life via mobile might in fact be more interesting than the life right in front of them. Stressful and inefficient, CPA is a never-ending effort not to miss anything, which can never be satisfactorily achieved.

Attempting to pay attention to everything at once can have significant social implications, many of them anti-social. Jennifer Aniston once famously dumped her boyfriend John Mayer for paying more attention to his mobile than to her. The actor Richard Griffiths, finding himself interrupted several times when performing on stage in London, turned to the audience and demanded: 'COULD THE PERSON WHOSE MOBILE PHONE IT IS PLEASE LEAVE? THE 750 PEOPLE HERE WOULD BE FULLY JUSTIFIED IN SUING YOU FOR RUINING THEIR AFTERNOON.'

These examples affect other people, but the worst consequence may well be for oneself. When a fashion stylist admits in a newspaper that 'I COULDN'T LIVE WITHOUT MY MOBILE PHONE', most right-minded people know it's time to pause and think harder about the way we handle technology. Do you have a balanced relationship with your phone?

fast and furious

YR MSGS R KLLG ME: THE TYRANNY OF TEXT

Have you ever regretted sending a text or email?

Texting has somewhat unusual technical origins. It was included as a throwaway feature on early mobile phones and none of the manufacturers thought many people would be bothered to manoeuvre their thumbs in all sorts of contorted patterns just to string a sentence together when it would be quicker to hit speed dial and talk. But the idea caught on in an unprecedented way. Americans send over 4 billion text messages a day, and Brits 1 billion. The volumes are phenomenal, and the technology keeps developing. Original SMS (Short Message Service) communications were restricted to simple ones such as 'HAPPY CHRISTMAS'. Now the recipient might receive a video clip or picture of a naked friend, or possibly a footballer.

The ubiquity of mobile devices and the rapidity of texting has generated a total change in social behaviour. In days gone by, a date meant specifying when and where in order to make the tryst work. 'THE DOG AND DUCK AT EIGHT ON TUESDAY' may well have been teed up several days in advance on a landline, and changing the arrangement was often out of the question on the grounds that the other person could not be contacted. Studies show that, having arranged such a rendezvous, the maximum amount of time anyone was prepared to wait was 12 minutes. This precision and reliability has been replaced with the infuriatingly vague 'I'LL TEXT YOU WHEN I'M IN THE AREA', leaving many a frustrated partner or mate twiddling their thumbs for hours, possibly sitting in the wrong venue.

Changing social behaviour is complicated. One twenty-something girl who was notoriously always 30 minutes late decided to mend her ways and start turning up on time, all to no avail because all her mates assumed she would always be late and so added half an hour to any proposed meeting time.

Many a relationship has ended in a vitriolic burst on SMS. It's fast and furious, and can be delivered with pretty much the same speed as a vicious insult to the face. And of course, phones do not come with a 'disengage when drunk' button. This might please young stags on a Saturday night when they 'text the ex' and get a positive response, but may be less appealing if insulting the boss and hugely regretting it later. Not that companies have a great track record in this area either – there have been many scandals involving text voting on TV programmes, and one company famously fired a large proportion of its staff using SMS.

Apart from social difficulties, obsessive texting can have more sinister implications. Many teenagers die every year when crossing the road looking at the message they are writing or receiving, rather than the traffic, and a number of government campaigns have highlighted these dangers. Worst of all, a woman was jailed in the UK in 2009 for hitting and killing a pedestrian while driving at 70 mph. She had been sending a text at the time. Do you text too much?

the tyranny of email

THOUSANDS OF EMAILS BUT NO CONVERSATIONS?

Do you check emails for several hours every day? They may be the brightest and the best, but they are not necessarily the nicest. That was the conclusion of a research study into the behaviour of college students in the USA. Apparently, their concern for other people's feelings is in sharp decline, leading to a GENERATION ME of self-centred, narcissistic and competitive individuals. The study concluded that the shift has been particularly noticeable since 2000, probably as a consequence of growing up with violent video games, online friend networks and an obsession with TV celebrities.

The tyranny of email keeps us in touch but also has the ability to drive us all apart. When 1.4 billion people send 247 billion emails every day, there is no doubt that we are EDICTED to chatting, but are we really saying anything? John Freeman argues that email encourages us to eschew face-to-face conversations with friends or colleagues in favour of the terse and anonymous immediacy of a computer-driven exchange. Psychologists note that users of modern technology are often driven by the same gambler's instinct that motivates someone to play a slot machine. You never know when something is going to land in your inbox so there is a tingle of excitement every time you check.

Using pen and paper provides one alternative. A US study showed that members of Congress paid very little attention to emails, classifying them on the same level as mass mailings and petitions. Personal letters and visits are regarded as much more powerful. And how about talking to people in person? We are becoming what psychologists call 'desocialized', losing the ability to take time away from our devices and communicate properly face to face.

As with any form of technology, email suffers tremendously from being used as the wrong medium. Placing the right message in the right medium is crucial to ensure that it is appropriate and effective. Many times people send what may well be the right message, but in the wrong medium. Any message that involves significant emotion, or severe news, is unlikely to suit email. Think 'I LOVE YOU', 'YOU'RE FIRED', or 'I HATE THIS'. Do you send or receive too many emails?

SLAVES TO CYBURBIA

DISTRACTION OR ACTION?

Do you find yourself distracted by social media sites when you should be doing something else?

In a study in 2007, Microsoft workers took an average of 15 minutes to return to what they were doing after they were interrupted by an incoming email or instant message. Once interrupted, they strayed off to reply to other messages or browse news, sports or entertainment websites. There is a huge difference between a planned and intentional action, and one taken as a result of a distraction. Many of us are now suffering from an ATTENTION DEFICIT SYNDROME – one in which we are so occupied dealing with the inbound that we are almost constantly off track. At any given moment, it is worth the individual knowing what they are concentrating on, and whether that is what they originally had in mind.

Slaves to CYBURBIA live in a virtual village, frittering away their lives sending messages across the ether, playing chess with people they will never meet, or dreaming up witty status updates on their profiles on Facebook. In Britain alone, 233 million man hours a month are spent on social networking sites. The number believed to be addicted to the internet in China is 10 million, and in 2010, 14 internet addicts fled the Huai'an Internet Addiction Treatment Centre in eastern Jiangsu province by tying up a guard and breaking out. That's real ediction for you.

The number of drivers using mobiles and hand-held devices continues to rise, despite the obvious dangers – they are supposed to being taking a specific action (driving) and are actively choosing a distraction. If you cannot resist answering your device as soon as it rings, you are likely to be suffering from **Distraction Overload**. And what's more, several studies show that interruptions impair creativity and memory. Individuals could do worse than pause to consider which is more important: what they originally intended to do, or the thing they were subsequently distracted by. Do you find yourself easily distracted by technology?

Everything

LOUDER

than

everything

else

EVERYTHING LOUDER THAN EVERYTHING ELSE

Have you ever failed to notice what someone was doing or saying because you were concentrating on your phone or computer?

Distraction is one thing, and may be momentary. INFORMATION ANXIETY, as identified by Richard Saul Wurman in the book of the same name, could be a more permanent state of affairs. As he famously showed, a weekday edition of the *New York Times* contains more information than the average person was likely to come across in a lifetime in seventeenth-century England.

Electricity reached a quarter of Americans 46 years after its introduction. Telephones took 35 years and television 26. It took broadband just 6 years.

It took two centuries to fill the shelves of the Library of Congress with more than 57 million manuscripts, 29 million books and 12 million photographs. Now the world generates the same amount of digital information 100 times a day. The scale and speed of these developments is mind-boggling, and the individual is entitled to ask whether they can cope with it if they feel that everything is sounding louder than everything else.

Technology is now so pervasive that Indians are more likely to have access to a mobile phone than a lavatory (563 million vs. 366 million). Meanwhile the Germans, who have something of a knack for finding a long word to describe a particular phenomenon, often suffer from FREIZEITSTRESSE – free time that makes you more stressed than work. In a switched-on world, they try to take time off but become more and more anxious when they are on holiday because they perpetually feel they are out of the loop. Do you sometimes feel overwhelmed by your technology?

the myth of
multitasking
the myth of
multitasking
the myth of
multitasking
the myth of
multitasking
the myth of
multitasking
the myth of
multitasking
the myth of

IT'S ALL THE (TECHNO) RAGE

Have you ever become deeply angry and frustrated by technology? Road rage is a well-documented phenomenon, and no one disputes that it exists because most people have seen it in action. Techno rage is a bit more obscure – often it happens in the privacy of someone's home, when they feel that they want to destroy their computer. Crashes and lost documents are major causes. Mobile devices can contribute to stress simply because they are the conduit of news, and as we all know, not all news is good news. Certain mobile ring tones have been found to symbolize and trigger stress, either because the listener has come to associate the noise with trouble, or because the ubiquity of the noise in public places represents a reminder of intrusion on privacy. One mobile phone manufacturer found both to be true of its most popular ring tone.

In a modern world, many think that they can handle it all at once, when in fact they probably can't. There must be physical and mental limits with regard to how much technological input one person can take, and we may have reached them. However, some scientists believe that our reliance on the web for information and communication might even be changing the way we think – literally. Browsing the web for extended periods is believed to affect our neural pathways, with implications for the way we respond to information and form memories. By skimming the surface of knowledge we cover more but absorb less, says one school of thought. If that is true, then exercising the mind is better achieved by concentrating on one thing at a time. But science remains divided on this.

The myth of multitasking has become a hot debate. Studies show that the most persistent multitaskers perform badly in a variety of tasks. They don't focus as well, they are more easily distractible, and they are weaker at shifting from one thing to another. In fact, they are worse at it than people who do *not* usually multitask. And yet having a lot on the go at once has become something of a badge of honour in its own right.

Pretty much everyone has experienced angry helplessness when technology breaks down. If we have not experienced it directly (although we probably have), we all know someone who has become distraught after leaving their mobile in a cab, failing to get a signal, losing a crucial document on a laptop, or running out of battery power somewhere inconvenient. It's modern life. We need a revolution to improve our ability to cope, and the resolutions in Part II should help. Do you get frustrated by technology?

'I just RANG to check if you'd got my TEXT about the EMAIL I sent you?'

HIDING BEHIND TECHNOLOGY

Have you ever used technology to avoid talking face to face?

As we saw in the last chapter, technology has played a part in what many believe to be a decline in social skills. Arguments for technology helping in this area are removal of social stigma, heightened awareness of what is happening beyond one's front door, the development of certain creative skills, and even the ability to find a suitable partner via internet dating. Almost every perceived benefit however, has a negative counterpoint that can be cited. Detractors point to reduced face-to-face interaction, anonymity that can lead to stalking, cybercrime, addictive gambling, online scams, online bullying, and a generation of users who struggle to interact successfully with the outside world.

Some people actively hide behind technology. The worse the news, the more likely it is to come via a computer or device. Looking someone dead in the eye and delivering bad news has never been easy. Now, if it's too painful to deliver in person, people can send a text, email, or instant message, or just leave a message. 'YOU'RE DUMPED' seems to be a case in point. More annoyingly for the recipient, multiple channels often lead to multiple messages – all about the same thing: 'I JUST RANG TO CHECK IF YOU'D GOT MY TEXT ABOUT THE EMAIL I SENT YOU?' Colleagues who want to cover their backs are major culprits here – everything seems to be delivered in triplicate to prove that it's your problem, not theirs. In a business context, this gets the issue out of your inbox and onto someone else's desk.

Technology can also provide scale to blur social distinctions. Three hundred 'friends' on Facebook may or may not mean what it says. Can the power of social media transcend the normal maths of how many relationships one person can realistically sustain? Some observers believe that such technology merely provides a new platform for bragging rights with regard to social popularity and business importance. Studies have shown that the primary purpose of the hand-held device is to increase the self-esteem of the owner. For underconfident people this may be no bad thing, but for those who are not short of bravado, or who lack subtlety, it could be unpleasant for all concerned. As Tom Peters has observed: 'IF YOU ARE CONSTANTLY ON YOUR BLACKBERRY, IT IS MOSTLY BECAUSE OF AN ARROGANT, CONSUMING SENSE OF SELF-IMPORTANCE TOTALLY DIVORCED FROM REALITY. THE WORLD WILL NOT COME TO AN END IF YOU ARE OUT OF TOUCH FOR 20 MINUTES. OR AN HOUR. OR A DAY.' Do you ever hide behind technology?

the cuddle chemical

CAN'T GET NO SATISFACTION (EVEN IN TOPLESS MEETINGS)

Have you ever communicated the same point using more than one method or device?

Which brings us to the fraught world of business. It is now customary for professionals to lay their devices on the table before a meeting like gunfighters with revolvers on card tables in saloons. This says to the other people in the room: 'I'M CONNECTED. I'M BUSY. I'M IMPORTANT. IF THIS MEETING DOESN'T HOLD MY INTEREST, I'VE GOT LOTS OF OTHER THINGS TO DO.' According to Jack Trout in *In Search of the Obvious,* this reduces many meetings to little more than gadget envy sessions. Recent research shows that a third of Yahoo staff regularly check email during meetings, and 20 per cent of them have been castigated for poor meeting etiquette in this regard.

Many businesses have realized the disadvantage of using email for selling. People now ignore email in the same way that they used to with unsolicited junk mail. Research shows that in any communication, body language accounts for 55 per cent of the power of your message, tone for 38 per cent, and words just 7 per cent. So while pinging off rapid messages using text, email or instant messaging increases speed and efficiency, it loses much of the passion, creativity and responsiveness that eye-to-eye or ear-to-ear contact can add. Removing the human can remove the humanity. Judicious use of social media can, however, yield tremendous benefits for companies, if they do it right. Many companies miss the point and try to hijack the online world for broadcast purposes. Peer endorsement and recommendation can be highly powerful, but it is not something that can be manipulated.

Human interaction releases a 'cuddle' chemical called oxytocin. We don't get this if we rely too much on virtual interaction and we risk isolation as a result. Even being present with colleagues while they are clearly not engaged because they are fiddling with some device or other isn't conducive to good work or productive relationships. In Silicon Valley, some tech firms became so exasperated by this problem that they tried introducing 'topless' meetings – as in laptop-less. Computers and BlackBerries were banned in the hope of making participants concentrate properly on the matter in hand. There was too much resistance and it didn't catch on.

If you have found yourself answering yes or maybe to many of the questions in this chapter, then you may well be EDICTED. Take the test to find out. Then we will move on to a set of ideas to help wean us off our technological drugs and lead a more fulfilling life. We are looking to restore the balance in your life, so that you can make technology work for you, and feel that you are getting the best out of it. At the end of each chapter you will find Taming Technology Tips – simple things that you can do to regain control.

NEED A REVOLUTION? TAKE THE TEST —

NEED A REVOLUTION? TAKE THE TEST

1. Do you ever interrupt a conversation with your partner or children to take a call?
   - ☐ Never
   - ☐ Occasionally
   - ☐ Off and on
   - ☐ Quite often
   - ☐ Always

2. Have you ever regretted sending a text or email?
   - ☐ Never
   - ☐ Occasionally
   - ☐ Off and on
   - ☐ Quite often
   - ☐ Always

3. Do you check emails for several hours every day?
   - ☐ Never
   - ☐ Occasionally
   - ☐ Off and on
   - ☐ Quite often
   - ☐ Always

4. Do you find yourself distracted by social media sites when you should be doing something else?
   - ☐ Never
   - ☐ Occasionally
   - ☐ Off and on
   - ☐ Quite often
   - ☐ Always

5 Have you ever failed to notice what someone was doing or saying because you were concentrating on your phone or computer?
- ☐ Never
- ☐ Occasionally
- ☐ Off and on
- ☐ Quite often
- ☐ Always

6 Have you ever become deeply angry and frustrated by technology?
- ☐ Never
- ☐ Occasionally
- ☐ Off and on
- ☐ Quite often
- ☐ Always

7 Have you ever used technology to avoid talking face to face?
- ☐ Never
- ☐ Occasionally
- ☐ Off and on
- ☐ Quite often
- ☐ Always

8 Have you ever communicated the same point using more than one method or device?
- ☐ Never
- ☐ Occasionally
- ☐ Off and on
- ☐ Quite often
- ☐ Always

9 Does your desire to stay in the loop distract you from the people who matter most?
- ☐ Never
- ☐ Occasionally
- ☐ Off and on
- ☐ Quite often
- ☐ Always

10 Have you ever been the victim of an online scam?
- ☐ Never
- ☐ Occasionally
- ☐ Off and on
- ☐ Quite often
- ☐ Always

▶ Scoring

Give yourself 5 for each 'Always' answer.

Give yourself 4 for each 'Quite often' answer.

Give yourself 3 for each 'Off and on' answer.

Give yourself 2 for each 'Occasionally' answer.

Give yourself 1 for each 'Never' answer.

▶ What your score means

If you scored 10–20, you are not edicted, and have superb willpower. Throw this book away, or give it to a friend who needs it.

If you scored 21–30, you are fairly under the cosh, so you should definitely read on.

If you scored 31–40, you are definitely edicted, but you can pull out of it. Read on, look hard at the Taming Technology Tips and start making some resolutions as soon as possible.

If you scored 41–50, it's amazing you've had the time to read this far, considering the amount of time you spend using technology. You are very edicted and definitely need a revolution. Try everything this book suggests, or call a doctor.

PART II: REVOLUTION RESOLUTIONS

03: THINKING

So it's time to tame your technology and get your life back. Whether you are totally ruled by machines or just feeling the tide turning against you, now is the time to act and regain control. It's time to confront the problem and make some resolutions that change things for the better. There are five main areas that require attention: thinking, talking, communicating, doing and being.

TOO MUCH RUBBISH AND NOT ENOUGH SENSE

THINKING, TALKING, COMMUNICATING, DOING AND BEING

Of course, they are all interrelated, but they are worth breaking down so that we can concentrate harder on what they mean and how they can affect our relationship with technology.

▶ Thinking

We don't do enough of it, even though it's completely free. Events overtake us. We need to rediscover the art of thinking clearly and use it to improve our quality of life.

Thinking takes time and requires a bit of peace and quiet. We can't think properly if we are distracted, and distraction comes in lots of ways. If you are trying to think and you are constantly interrupted by inbound alerts, you may never crack the problem. So it doesn't work if the phone keeps going, or an email, text or instant message comes in, or if we are trying to multitask and not making a good job of it. So we need to create the time and space to get away from technology at certain times to think properly.

▶ Talking

We do too much of it, often without having thought first. We talk too much rubbish, and not enough sense. It's time for a new, more considered approach that reflects what we feel more accurately and makes it easier for others to understand us.

If we don't think, then there is a very strong chance that the quality of our talk will suffer. This works in two directions. Firstly, people who don't think first tend to talk nonsense in any medium. Secondly, the person who has to listen to the resulting drivel is either frustrated or hasn't a clue what the other person is talking about. Neither state of affairs is good. People who blurt out any old rubbish without thinking create the impression that they don't think much (which would be true), and are also lousy communicators, so nobody wins. Many claim that they can't work out what they think until they talk it through. This may be true, but at least they should have the courtesy to alert the other person that that is what they are about to do. The recipient of the outburst (the wafflee) has the right to know what they are in for (from the waffler). Talking straight affects how well you can communicate through your technology.

▶ **Communicating**

We have so many methods of communicating available to us that we frequently choose the wrong one. We need the right medium for the right message, and a clear understanding of the suitability of each for the task.

Communicating is a broader skill. It includes talking of course, but also embraces many other media, such as the phone, email, text, instant messaging, social media, presentations, and much more. Communication is two-way, and can rarely be achieved by a monologue, unless you happen to be the president making a speech, and even then there is plenty of room for misunderstanding. This means that listening is a crucial part of communicating if it is to be successful. Choosing the right medium is absolutely critical. How many times have you chosen the wrong medium for the message? It happens all the time, and the more technological options we have, the more baffling it becomes.

▶ Doing

We do far too much of the wrong stuff, which often means we do not have enough time left for the right stuff. We need to be able to distinguish between what matters and what doesn't to make better use of our time.

Doing is the action part. Without it, nothing happens. Writers do not sit waiting for inspiration to strike. They sit and write until something emerges. Anyone can have a great idea, but it is worth nothing unless it gets done. In that respect, execution has become one of the most valuable traits in the modern world. People mistakenly think of execution as the tactical side or menial side of things. They couldn't be more wrong. People who get seduced into concentrating on the so-called bigger picture usually fail to deliver. In 2000, 40 of the top Fortune 200 company chief executives were dismissed because their organizations had failed to do what they promised. There should never be a gap between thinking and doing. Technology can fool people into thinking that something is being done, when in fact it isn't, as we shall see.

▶ Being

We can all be better. We should define our own characters, not allow them to be defined by our possessions or technology. Having a better life starts with knowing what you are all about, and emanating that style.

And finally, being. Existence, and the quality of it, is the very essence of life. So it must be worth making sure that what we are suits us well. Our self-determination and self-esteem may well be enhanced by technology, but they certainly should not be ruled by it. Getting your attitude to your technology right and having a decent balance leads to less stress and a more fulfilled life. It starts with good thinking, emerges as well-considered talking and communicating, and manifests itself as doing, which leads to high-quality being. And if you believe the words of Rene Descartes – 'I THINK, THEREFORE I AM' – then the cycle from thinking to being is complete.

THE ESSENCE OF A DECENT LIFE

MIND GAMES

Thinking is the essence of a decent life. The more people do it, the more fun and fulfilment they have. Everything that gives us pleasure involves a thought – generous deeds, great design, fantastic entertainment, everything. Someone, somewhere, thought carefully about how something could be, and then they crafted it and made it exist. We can't all be inventors and master craftsmen, but we can pause for a while and make our own little sphere of influence that little bit better. Thinking is how we bear responsibility. Without it, we don't take any, and so we relinquish what happens to us to other people. That's no way to live. If you don't pause to think, you'll be dictated to by everybody else who does. And if you don't pause to think about your technology, then you'll be dictated to by your machines. Life suddenly becomes interesting when you take charge. As the philosopher John Stuart Mill said, 'ONE PERSON WITH BELIEF IS EQUAL TO A FORCE OF 99 WHO HAVE ONLY INTERESTS.'

Stimulating thinkers the world over generate the kind of competition that we all cherish and enjoy. Great debating, great sportsmen and women, great chess players – they are all brilliant thinkers. The internet has generated a much bigger forum for the exchange of ideas and views, but it has also created a paradox of choice, as we saw in Part I. Thinking needs to be done in a stimulating environment that suits the thinker, and we can replicate our own modest version of this quality all on our own. It takes some discipline, but it's not really that hard if you concentrate and put your mind to it. Let's break this down into small chunks to make it clear, and probably less daunting.

Starting with the minuscule, would you ever walk into a room without deciding why? Or drink a glass of water without deciding you were thirsty? No you wouldn't. Each action involves an initial thought, however seemingly small or inconsequential. In the next 5 minutes, you will make many such decisions, and when you add them all up, you'll have filled a day. So, to make today more fruitful, take a minimum of 5 minutes, but preferably 15, to decide how you are going to interact with your technology today, and why that is appropriate. Ask yourself: Why am I writing this email? Why am I logging onto this social network? Why am I using this machine? If you cannot answer why to your own satisfaction, then there may be no value in using it for doing that particular thing.

It is then a simple matter to extend this principle to a month, a year, or a lifetime, but of course the longer the time span under consideration, the longer you will need. This is not an anti-technology stance. In fact, technology can enable our ability to think if used thoughtfully. As Erik Qualman reports in *Socialnomics,* an 83-year-old American called Bill Tily has made a habit of printing out his social media updates and highlighting those that are not contributing to what he calls a full life. He then ceases engaging in 'unfruitful activities' from then on. We don't do enough thinking, even though it's completely free. Events will always try to overtake us, but we urgently need to rediscover the art of thinking clearly and use it to improve our quality of life.

Thinking is free

THINKING IS FREE, SO DO IT MORE OFTEN

How many times have we heard someone say that they haven't had time to think? Millions of people say it every day, but what does it actually mean? If you analyse the phrase carefully, it is complete nonsense. Every sentient being spends the entire day thinking, absorbing circumstances and reacting to them. Of course, the phrase is not literal. What it really means is that they haven't had time to pause and think about the things that really matter, because lots of irrelevant stuff has got in the way. This is a tragedy, and it is your job to create the appropriate time to rectify the position.

Why is this so important? Because, although you may claim that you are too busy to create the time, if you haven't worked out whether what you are doing with your technology is the right thing, then you may only be busy pursuing all the wrong things. Never engage with it unless you know why you are doing it. This sounds blindingly obvious, and yet people frequently do.

So now is the time to get thinking. You need to set aside the time and create the appropriate conditions. Some people like total peace and seclusion, others like something to shake them up. Work out your style by considering whether you are more likely to have some decent ideas if you sit on top of a mountain, have a massage, get on the running machine, disappear to a country cottage, drink a bottle of quality wine, leave the country for the day, visit the zoo or go fishing. The activity or circumstance doesn't matter, so long as it is different from where you normally are, and what you normally do.

Remove yourself from the persistent attention of your technology, even for an hour, and think. You have to enter the thinking process in the right frame of mind. It's no use being petrified, depressed, cynical, paranoid, resentful, jaded, or any other negative emotion. It is okay to be a bit vexed or concerned. It is all right to be mildly sceptical. It is fine to be quizzical. In fact, that should positively be encouraged. Your objective should be to reflect on how you interact with your machines and view it as though you were someone else looking at you. Strange, and quite detaching, but ultimately rewarding. Write down how long you spend on machines each day. Keep a diary for a week. Survey the results. How do you fare?

GOD IS IN THE
DETAILS

SMALL THOUGHTS ARE AS VALUABLE AS BIG ONES

A lot of people get hung up on planning the 'next big thing'. But who is to say that the next big thing has to be big? In technological terms it could be quite small, such as moving your laptop to a different place, or adjusting a setting on your mobile. Sometimes tiny increments of change make amazing things happen. As Malcolm Gladwell says in *The Tipping Point,* little things can make a big difference if cunningly applied. So don't panic about the fear that you need to come up with something outstandingly original. People rarely do. Occasionally someone like Edison will invent a light bulb, but that's a bit beyond our remit.

The architect and furniture designer Ludwig Mies Van Der Rohe once proclaimed that 'GOD IS IN THE DETAILS'. There is often great mileage to be had from lots of little ideas. Little ideas are great. They are less hard to come up with, they are usually cheaper and easier to implement, and they can be done more quickly than something big and scary. This enables you to work out rapidly whether they are any good or not. Little ideas can be test-driven constantly, refined, enlarged, developed or withdrawn with the minimum of fuss. Think hard about your relationship with your technology and try making your next big thing small. You might surprise yourself.

time for a re-think

THINKING OR JUST REARRANGING YOUR PREJUDICES?

William James said that a great many people think they are thinking, when they are merely rearranging their prejudices. If everything is too samey, or things aren't going that well, it's time for a re-think. And that does not mean rearranging your prejudices, or dreaming up new reasons to prove that you are right about something. It means taking a hard look at what you've got and working out whether it is any good or not, and whether you like your circumstances. If you have any doubts about your approach to technology, it has to be done. Even in the unlikely event that you don't have any concerns at all, it is still a great thing to do. Everything can always be made better or more stimulating. The philosopher Bertrand Russell famously said: '**MANY PEOPLE WOULD SOONER DIE THAN THINK. IN FACT THEY DO.**' Make sure you don't die, if only technologically, from a lack of thought. Work hard to clarify your thinking and don't shy from the task. It's your life, and you want it back.

don't try to **fix everything** at once

HOW TO OUT-THINK YOURSELF

If you want to out-think yourself, you first need to face up to your failings. The purpose of this is not to beat yourself up, but to help you recognize what you tend *not* to do well in relation to your machines. You can then set about working out ways to work around those deficiencies. Mostly, these are just small shortcomings that are mildly irritating, but if you have several deficiencies they can put some fairly large barriers in the way of you thinking clearly. Knowing yourself helps to address problems and think more productively.

Out-thinking yourself requires that you use your calm, controlled moments in order to anticipate what you will get wrong or fail to do at a later uncontrolled moment. If you often forget your house keys, then introduce some changes to counteract the problem. It only takes a minute to dream up some initiatives that might help: put a note on the back of the door to remind yourself, hang them on a hook there, always put them in the same place, keep a spare set at work or with a neighbour, put them in your trousers, briefcase or handbag the night before, and so on.

If you often run out of petrol, you can do the same. Put a note on your steering wheel saying petrol, always fill up on the same day of the week, always fill up when the tank is half full, put a spare can in the boot, or buy a bicycle. The knack is to admit at the outset that you will probably not do the job, and then work out the most effective way to make sure that you do. Take one thing at a time and don't try to fix everything at once. The Taming Technology Tips give you lots of ideas about how to apply this to your technology.

taking responsibility

YOUR LOCUS OF CONTROL

Grabbing control of this whole area and taking responsibility for your actions is what psychologists call your locus of control. Julian Rotter's Social Learning Theory suggested that there are essentially two types of people, depending on their upbringing. Those with an internal locus of control believe that reinforcement depends on personal efforts. They think they are in charge of their lives and act accordingly. They are physically and mentally healthier and more socially skilled. Their parents tend to have been supportive, generous with praise, consistent with discipline, and non-authoritarian. Those with an external locus of control believe that reinforcement depends on outside sources – so they make fewer attempts to improve their lives and get things done.

Clearly, you cannot change your upbringing, but you can aspire to the qualities represented by an internal locus of control. That means taking responsibility for your technology, recognizing that improvements can be made, and acknowledging that it is down to you. It's no good sitting passively claiming that there is nothing you can do, because there is. As Jim Watson points out, 'YOU CAN SAY, "GEE, YOUR LIFE MUST BE PRETTY BLEAK IF YOU DON'T THINK THERE'S A PURPOSE." BUT I'M ANTICIPATING HAVING A GOOD LUNCH.'

EMBRACE THE ART OF INQUISITION

TURN ON OR TURN OFF?

If the philosopher Bertrand Russell was right that many people would sooner die than think, then we can reasonably assume that a lot of people simply don't like doing it. Not only is this a shame, but also it is totally counterproductive for any individual. Absence of thought effectively cedes decision-making to the random events of life. That abdication of responsibility may well allow the individual to claim that they are a victim of circumstances, but they are wrong to believe it. Being the passive recipient of everyone else's actions is no way to live. As George Bernard Shaw pointed out, progress in the world relies solely on the unreasonable man, who shapes the world rather than allows it to shape him. This is how you should approach your relationship with your technology. Only then will you get your life back.

Thinking needs to become a turn on, not a turn off. Where technology is concerned, there are two ways to approach this. The first is to deliberately choose a new and stimulating online environment to help your thinking, and the second is to turn it all off. Choose the one you prefer, or try both. Then when you have the setting you want, you can begin to concentrate on the matter in hand, and, given the frenzied pace of modern life, there may well be many issues to review. The brain is a bundle of bodily tissue and needs exercise like any other. If you don't use it very often, it will hurt like hell when you do. As Barbara Castle once commented:

'IF A PROBLEM IS HARD, THINK, THINK, THEN THINK AGAIN. IT WILL

HURT AT FIRST, BUT YOU'LL GET USED TO IT.' When did you last give yours a really good work out?

The best type of thinking is simple, not complex. Begin with the sort of questions that children might ask. Ask 'why?' three times in a row to extract a better reason for doing, or not doing, something. Draw up lists of questions and work through them carefully. If you really don't know the answers, then do some research. Learning is discovery. Embrace the art of inquisition. Think about what to say, and what *not* to say. Think about which technological medium is the most appropriate in which to say it, so that you can express your feelings in a helpful and effective way. It's time to make some resolutions to start your revolution. Look back at how you fared in the test and consider how much you need to think about your approach to technology. Now think again based on this chapter, and consider whether any of the tips can help.

Taming Technology Tips for Thinking

TAMING TECHNOLOGY TIPS FOR THINKING

▶ **1 Turn off every device and appliance you have to create thinking time**

This may sound a bit extreme but it's a great discipline, even if you only ever do it once. You'll probably conclude that there is at least one piece of machinery that you never use anyway, and you'll probably reduce your electricity bill.

▶ **2 Walk away from your machines and go somewhere quiet**

If you really can't unplug the stuff, perhaps if you are at work, then hide yourself away somewhere appropriate. Don't tell anybody where you are. Enjoy the silence and really think. It will hurt at first, but you'll get used to it. Do not take your mobile phone with you – it is not your Siamese twin. The world will not fall apart if you cannot be contacted for a short while, but *you* might fall apart if you can be reached every moment of the day.

▶ **3 Try creating some online thinking time**

Choose a totally new online environment – one that you never normally use or engage with. This could be a new world of music, film, photography or writing. Use that stimulus to push your ideas to new places.

▶ **4 Use something simple like a pen and paper to note any good thoughts**

It's refreshing to use a different medium, particularly if you spend a lot of time tapping away on a keyboard. You might even sketch some pictures or diagrams. You will probably also find that you express your ideas differently.

▶ **5 Review your day, week, month, or year**

The longer the period in question, the more thinking time you will need. Choose your medium and your time period. Think about what you feel you have to do and compare it with what you really *want* to do. If there is a significant difference between the two, work out what needs to change, pledge to change it, and write it down.

▶ **6 Move your technology to give you more peace**

Your approach to your technology could change depending on where it is. If you are constantly tempted to log on or generally fiddle, then move that device into another room and keep it there. For mobile devices, put them as far away as possible when you need to think. Get into a habit that forces you to actively decide to interact with a machine, rather than have it interrupt you.

▶ **7 Remove alerts on all computers and devices**

If you react like Pavlov's dog whenever your computer or hand-held device makes a noise, then turn these alerts off. *You* decide when and for what period of time you wish to interact with them, not the other way round. Anyone with Obsessive Mobile Disorder needs to break the pattern. Don't be a lapdog to your laptop.

▶ **8 Always turn your phone off when you want to think**

Honestly, you can wait a few minutes for a message, and bear in mind that seeing or hearing what the other person has to say, and having an extra moment to think about it, may alter or improve your response. So you will also gain another type of thinking time.

▶ **9 Design a regime for the only times you will look at your devices**

Once in the morning, once at lunchtime, and once at the end of the day. That should be enough. If you feel it isn't, then do ask yourself when you are going to get anything done. Carve the day up into chunks, and choose when you will do things on *your* terms, rather than simply deal with everyone else's requests.

▶ **10 Make thinking time a part of your routine**

Once you are sold on the idea, thinking should become an integral part of your day, week or month. Grab your diary now and map out times when you actively want to think. Fifteen minutes each morning, an hour at the beginning or end of each week, a day a month, or a week a year are suggested amounts for you to consider. It's time to regain control.

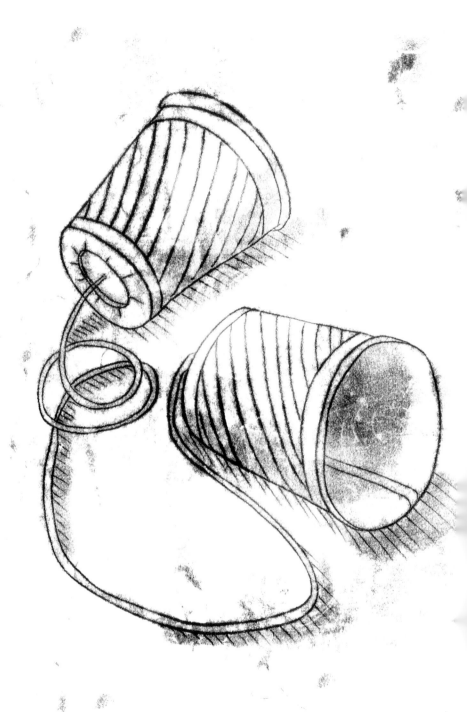

04: TALKING

brevity =equals= intelligence

THE ART OF BRACHYLOGY

BRACHYLOGY. It's an odd word but one really worth considering. It means a concise style in speech or writing. Making something pithy is far harder than rambling on, as all wafflers know. As Blaise Pascal, the scientist and philosopher who lived in the 1600s, once pointed out, 'I HAVE MADE THIS LONGER THAN USUAL BECAUSE I DID NOT HAVE TIME TO MAKE IT SHORTER.' (This saying is sometimes attributed to Mark Twain, but Pascal certainly came first.) It sounds contradictory, but it isn't. Anyone can drone on for a long time, but thoughtful people think about what they are going to say first, then they say it. And when they do, it's usually worth listening to. So let's look at how we can talk more succinctly with our machines, and so free up time to work more effectively with them, or do something else.

Brevity equals intelligence. The prevailing mood of much modern speech is long-winded, circuitous and not necessarily very well thought out. You only have to listen to a politician to hear that. You should always aim to talk straight – to construct clear, jargon-free sentences and say them out loud. Many find this difficult, but with concentration it should be within everybody's grasp. The clarity of your spoken approach will be a direct reflection of your clarity of thought, and will have a huge bearing on your use of, and power over, your technology. Using speech correctly should reduce the amount of time you spend using machines, especially phones.

The less time it takes to articulate a point, the better expressed it is. The more you leave out, the closer you get to the heart of the matter. Your first instinct of what to say may well include all sorts of broad material, but it is not until careful thought is applied to the main point that much of it can be discarded. More material does not necessarily strengthen an argument or improve clarity. In some complex academic and technical areas it might, but in most areas of life, it doesn't. Skilful editing and the ability to filter out extraneous material is a crucial asset for anyone who wants to tame their technology effectively.

You will probably have heard the expression that less is more. This is the notion that simplicity and clarity lead to good design. It is a phrase often associated with Ludwig Mies Van Der Rohe, one of the founders of modern architecture and a proponent of simplicity of style. Less really *is* more. We talk too much, often without having thought first. We talk too much rubbish, and not enough sense. It's time for a new, more considered approach that reflects what we feel more accurately and makes it easier for others to understand us. Brachylogy is the key to using technology.

IT STARTS IN YOUR HEAD

THINKING AND TALKING STRAIGHT

The link between thinking straight and talking straight needs to be established in your mind. Pay attention to the connection, and consider the consequences of talking twaddle. You can't talk straight if you can't think straight, and translating straight thinking into straight talking pays tremendous dividends. The pitfalls shouldn't need articulating. If a person doesn't talk straight socially, then other friends cannot react specifically. If a partner doesn't talk straight, then the other partner will most likely miss the point. If a parent doesn't talk straight, the child will be confused. If the child doesn't talk straight, then the parent is none the wiser. If a boss doesn't talk straight, then subordinates cannot take suitable action. If staff don't talk straight, then the boss is misinformed and cannot take appropriate action. It's a nasty chain of confusion that usually triggers the need for further action unnecessarily.

Talking straight is a vital component of clear communication, and crucial to the speed and efficacy with which things get done. This has an immediate bearing on whether you will then have to interact with technology or not. If a thought, request, or instruction is unclear, then there is a strong chance that there will have to be another conversation, and that probably means another phone call, email, text, or instant message. Or several of these. Or a mixture of all of them. In a work context, this could get even worse, since it may well generate the need for a re-draft, or a whole new proposal. It all means extra work, and more time on a computer, with a phone nailed to your ear, or fiddling with a device. It's better to get it right first time.

The best communicators understand how language works, and concentrate hard on it. The great thing about concentrating on how language works is that you gain two fundamental benefits:

1 You can talk straight yourself.

2 You can identify when someone else is not talking straight.

Guidance on how to talk straight is all around us. Good books, spoken radio shows, great speakers, and, of course, the dictionary, are all sources of learning and should be consulted frequently. Inquisitive students of language should try to look at a dictionary every day. Try to understand the correct definitions of figures of speech and use them appropriately. When used in the right context, they can enhance communication and understanding, and enrich your life. If people understand you better, then there will be less confusion, and therefore less stress. It starts in your head, and it's your job to explain it to the rest of the world clearly via your machines.

talking
rubbish
to
yourself

SPOTTING WAFFLE

There are two types of waffle: internal and external. Internal waffle is talking rubbish to yourself. It sounds near impossible but millions of people do it. They can just hear the words, unspoken, rattling around in their heads as they embark on a marathon bout of self-delusion. It's insidious stuff, and arguably even more potent when unsaid than when uttered out loud. Internal wafflers believe their own propaganda and base decisions on it – often important ones. They then regurgitate their muddled thinking through their technology as observations and requests to others, usually causing chaos and confusion in the process, and generating the need for more and more exchanges.

It is beyond doubt that we live in a tricky verbal world. What can you do about it? Here is a ten-point guide to spotting deceptive waffle. Remember: a lot of the language is very hard to penetrate, so you need to listen hard and question very deeply. Check any statement or request for action for the following:

▶ 1 False arguments

If you know your facts and the other person is incorrect, then you should either ignore them or refuse to act on their instructions. In fact, you may not even be able to. This can also apply to people who change their minds without any justification. They will often say the opposite of what they said previously, without acknowledging that they have changed their stance. It is your job to remind them of this, and provide proof if necessary. Without being overly antagonistic, they should at very least be made to admit that they have changed their mind, and explain why.

▶ 2 Circular arguments
If the person in question keeps returning to the same point, then there may only be one point. If it's valid, then fine. If it's not, ignore it or do not act on it.

▶ 3 Repetition
Similar to the circular argument, but not quite the same. In the circular example, the speaker will head off elsewhere and then return. In the case of repetition, they just keep saying the same thing, albeit using different words.

▶ 4 Incorrect conclusions
These are 'synapse jumps' and what the Romans called a non sequitur. The synapse is the point at which a nerve impulse is relayed from the terminal portion of an axon to the dendrites of an adjacent neuron. The charge jumps the gap to let your brain know what's going on. People often jump a gap and end up in completely the wrong place. A non sequitur is just that: a statement that has little or no relevance to what preceded it – literally: 'IT DOES NOT FOLLOW'.

▶ 5 Spurious sources
Some people make themselves sound authentic by quoting sources, but they may not be valid. Check these assertions carefully and question them closely.

▶ 6 Irrelevance

It is extraordinary the amount of irrelevant material that is wheeled out to validate a point. In fact, the more evidence is offered, the more inquisitive you might wish to be – the 'THOU DOTH PROTEST TOO MUCH' syndrome.

▶ 7 Weak points

Most decent cases either argue for themselves or have one or two solid support points. If someone puts up a chart with support points, the weak points at the bottom always undermine the strong points at the beginning. If this is the case, review the entire argument.

▶ 8 Cliché and jargon

Don't trust it. It is usually disguising something.

▶ 9 Inconsistency

Inconsistency is to be viewed suspiciously if it appears in the same burst of speech. But also keep an eye out for it over time. Many people gainsay themselves in subsequent statements.

▶ 10 Vagueness

Vagueness should be easy to spot. If the person can't come to the point, or clearly hasn't got one, then ignore them or don't do what they say.

By learning to spot all this, and by not doing it yourself, you will automatically be talking straight, and so reducing the amount of repeat tasks that will force you to engage with technology unnecessarily.

'HAVE YOU GOT A MINUTE?'

THE SPONTANEOUS WORD DUMP

One of the worst examples of bad talking is the spontaneous word dump. This takes many forms. Any of your machines could burst into life at any time and suddenly there someone is, talking to you. It could be your phone, in which case the caller may or may not ask whether this is a convenient time for you, and in our time-pressed world, it very rarely is. It might be a text, an instant message or an email. All were originally designed for brief notifications or exchanges of information, but have now morphed into just more media for extended diatribes. Or you may be at work and hear the dreaded words: '**HAVE YOU GOT A MINUTE?**' Whether you are at your desk concentrating on a task, or on your way to somewhere and subject to the 'corridor surprise', chances are you *haven't* got a minute, but the person is going to start talking anyway.

The spontaneous word dump is one of the most thoughtless acts a human being can perpetrate on another. It gives the other person no warning, pays no regard to whether it is convenient for them, and is totally selfish. It is intended solely to benefit the dumper, not the dumpee. Regular dumpers use more and more cunning ways to dump, and technology is their unwitting accomplice. People who receive too many phone calls eventually stop answering. People who are intercepted in the corridor one too many times can barely make it to the washrooms in time. One beleaguered executive complained that whenever she left her desk, a phalanx of hangers-on and questioners used to form behind her as though she were the Pied Piper of Hamelin. If you are a spontaneous dumper, it is time to take stock and realize what a monumental waste of time it is. No one wins if everybody is speaking before they think, except possibly the phone companies.

Listening
is a dying art

HOW TO LISTEN PROPERLY

Conversations should be two-way, so they should be equal. This vests responsibility in both parties. It is the duty of the speaker to speak clearly and concisely having thought carefully about what they wish to say, and it is the role of the other person to listen properly. Listening is a dying art. It has been said that there are no conversations any more, just intersecting monologues. Martin Lomasney pointed out that you should: 'NEVER WRITE WHEN YOU CAN SPEAK. NEVER SPEAK WHEN YOU CAN NOD.' Listening is a skill. Ernest Hemingway said: 'I LIKE TO LISTEN. I HAVE LEARNED A GREAT DEAL FROM LISTENING CAREFULLY. MOST PEOPLE NEVER LISTEN.' And Doug Larson said that wisdom is the reward you get for a lifetime of listening when you'd have preferred to talk.

The greatest gift you can give another person is the purity of your attention. So next time someone talks to you, listen properly, particularly if you are using aural technology. Start by saying nothing. Let the words roll over you, take it in and try to understand. Then when there is a pause, offer your opinion. There is a catch though. If they are a waffler, then this approach might drive you mad. As we have seen, a conversation bears a two-way responsibility. If you are the regular recipient of a deluge of monologue, then you may need to have a word with the habitual waffler, or dream up some containment strategies. These can be devious or pragmatic, and need to be judged depending on the context. Getting away from the source of the waffle may be essential. Here are some suggestions for escape lines:

- 'I am late for a meeting. Can we pursue this later?'
- 'I am right on a deadline. Can we pursue this later?'
- 'I've just remembered something important. Can we pursue this later?'
- 'Can you talk to a colleague about it?'
- 'Can I talk to a colleague about it?'
- 'Will you excuse me? I need to have a quick word with someone.'
- 'Will you excuse me for a moment? I need to go to the bathroom.'

HOW TO SAY 'NO' POLITELY

Another use of talk to get your life back is to say 'no' more often. It sounds a bit blunt but it can be done politely as we shall see. These suggestions have to be handled sensitively. Your ability to use them will depend hugely on your level of experience, the quality of your relationship with the person you are talking to, and the tone you adopt. There are lots of ways to say no, but it is important to realize that aggression never works. Genuine concern to get it right usually does. This is particularly important when dealing with those in authority, or those with precious egos. The knack is to depersonalize the conversation and concentrate on the issue, not the person. Bluntly refusing to do something without offering any alternative isn't going to get you very far. But you should become much more inquisitive so that you don't blindly accept whatever you are told to do without believing it to be a good and useful action. Here are some ways to say no politely that can be applied through a variety of machines.

'THE CONSEQUENCES OF DOING WHAT YOU SUGGEST ARE X AND Y [SOMETHING NOT GOOD OR HELPFUL]. ARE YOU STILL SURE ABOUT THIS?' Let them consider and give them time to think. They may well withdraw the idea once they have heard your opinion. This can be particularly helpful if the person in question respects your opinion.

'I REALLY WOULDN'T RECOMMEND THIS. IT GOES AGAINST ALL MY PREVIOUS EXPERIENCE.' Obviously you need to be reasonably experienced to pull this one off, but it does work when you know you are respected by the other person, and is particularly important if you are part of a team.

'I REALLY DON'T AGREE BECAUSE OF REASONS X, Y, Z.' This is a purely rational response, and can be deployed when you have a good understanding of a subject and all the issues. Try to keep it unemotional, and let the facts speak for themselves. If they are sufficiently strong, then you won't have to push the point.

'MAY I DISCUSS THIS WITH SOMEONE ELSE AND CALL YOU BACK?' This is admittedly a case of stalling for time, but it is valid. Many people who ask for something either forget or change their mind by the following day. By building in thinking time, you take the edge off confrontation and allow for a more considered response later.

'I HAVE ALREADY DISCUSSED THAT POSSIBILITY AND REJECTED IT AS INAPPROPRIATE FOR THE FOLLOWING REASONS.' A lot of people are very inconsistent. Sometimes a new request flies in the face of what has been discussed before, and it is often appropriate to remind them of important discussions that have already taken place. Recap on any previous conversations, meetings and discussions that have a bearing on the outcome.

'I WANT THIS TO BE REALLY CAREFULLY THOUGHT THROUGH. CAN I THINK ABOUT IT AND DISCUSS IT WITH YOU TOMORROW?' In many walks of life, a solution that takes longer is valued more, whereas the immediate response is regarded as too spontaneous to be well considered. If this is true in a particular circumstance, then you need to understand the value of retiring to think. It also buys you some time and gives you the chance to seek advice if you want to.

'IT WOULD BE REALLY HELPFUL TO UNDERSTAND WHAT HAS CHANGED BECAUSE WE AGREED THIS YESTERDAY.' It is amazing the number of times when you think something is agreed and then suddenly it isn't. Even stranger, the person delivering the news often makes no mention of what has changed and what the reasons are. In these circumstances, it is your job to ask the question. It is only reasonable that, if you are being asked to do something new, you should be told why. If you disapprove of the new direction, then say: 'MMM, THIS SHEDS A DIFFERENT LIGHT ON THINGS. MAY I HAVE TIME TO CONSIDER?'

'I CAN'T DO ALL OF THAT, BUT I MIGHT BE ABLE TO HELP WITH SOME OF IT.' Split the problem into parts and offer to solve some of them, but not all of them. This approach allows you to be helpful and responsive, without taking on masses of extra things to do. It also gives you the chance to cherry pick the bits that you are best at or enjoy most.

There are also plenty of rather more devious approaches to saying 'no'. I don't necessarily approve but they can be amusing. Here are my favourites:

▶ Confuse the issue by continually introducing new considerations until they have forgotten their original request.
▶ Say 'yes'. Then find a way of saying no later.

enrich
your
life

TIME FOR A NEW LANGUAGE

It's time for a new language – one that understands the link between the brain and the tongue; one that respects the engagement of the one before the deployment of the other; and one that makes best use of what technology can facilitate. Einstein famously remarked that you do not really understand something until you can explain it to your grandmother, and it's true. The best way to determine whether you are talking straight is to try it out on your mum, your grandmother or a mate in the pub. If they don't understand what you are talking about, then it's too complicated, inconsistent, or simply not expressed well enough. And it will be even more prone to misunderstanding over the phone, text, instant message or email.

The mate, mum, or grandmother test can also work remotely. Even if they can't be there in person, imagine whether they would know what you were talking about. Would they understand it? If no, then change it. Does it sound daft? If so, then change it. Do you feel stupid when you hear the words out loud? If yes, then change it. Practise regularly how to speak concisely and clearly.

If you want to tame technology and get your life back, talking properly provides a crucial bridge between thinking and communicating. It is a form of communication in its own right, and determines whether you need to use another medium. If you can use words to achieve the job on hand, then do it. That's your first port of call. If it were possible to ditch every email and phone call, and conduct all your affairs face to face, then most people would. That's not how the world works, but it is the place to start. So before you dive into another round of emails, consider whether you could achieve what you want by chatting to people instead.

Talk takes many forms. If you can't talk directly, then consider meeting up if it's not immediately urgent. If not, then calling is better than email, so put in the effort. A possible exception to this might be when personal interaction is limited, in which case there may be a need to record thoughts and actions on email, so long as this does not become habitual or result in a backside covering mentality. Video conferencing and webinars are also preferable to email, because at least they can convey the body language and eye contact that play such a large part in effective communication. Your voice contains subtlety of tone and nuance that the written word never can. That's the sequence: talk in person, replicate the conditions of a face-to-face meeting using technology, speak on the phone, then only email, text or instant message as a last resort.

If you follow this simple sequence, then a number of helpful things will happen. Firstly, you will have more proper direct conversations with people, which will enrich your life, whether socially or at work. You will learn more, enhance your relationships, and smooth out many issues more easily and quickly. Crucially, do note that a face-to-face conversation involves no technology whatsoever. Secondly, you will make and receive fewer phone calls, because more of your conversations will be direct and personal. That will reduce the number of hours you spend on the phone, and the size of your bill. Thirdly, you will send and receive fewer emails, texts and instant messages. You will also become better at talking.

Taming technology tips for talking

TAMING TECHNOLOGY TIPS FOR TALKING

▶ 1 Don't start talking until you have finished thinking

Think at the beginning of the day about what you will be discussing, and work out what you think before you dive in. Concentrate on conversations and subject matter, not on types of technology that can transmit information. If you are caught on the hop, take a second or two to think before you start answering, or ask for time to consider.

▶ 2 Talk straight, based on straight thinking

Consider how you are going to phrase things. Work out a logic chain that explains how you got from *a* to *b* and arrived at your point of view. Rehearse how you are going to start and finish what you are going to say, and which technology to use.

▶ 3 Be concise, don't waffle, and stick to the point

If you are a self-confessed waffler, then make every effort to stop doing it. Fillet your point of view before opening your mouth. Look for synapse jumps that don't make sense, and repair the disconnects. Say your piece, then shut up.

▶ 4 Have the confidence to believe that brevity equals intelligence

The shorter the better. The more you have thought about it, the less time it will take to say. Don't protest too much. Intelligent, confident people can express things fast and clearly, without any unnecessary baggage.

▶ 5 Learn how to spot waffle and develop skills to deal with it

If you are the wafflee on the receiving end, work hard on coping strategies. Regard it as your job to educate the waffler and control their ramblings. Conversations are two-way, and you need to live up to your responsibilities.

▶ 6 Don't generate spontaneous word dumps

Never ambush someone else with a random piece of nonsense, whether using technology or in person. It's rude and unproductive. Curb your instinct to blurt things out. Pause and come up with something more measured. You will increase your confidence and come across as more intelligent.

▶ 7 Work hard to develop your listening skills

It's harder than it sounds but it's worth it. Try saying nothing for a while – you might learn something. The wisest, best-informed people listen the most. In particular, practise this on the phone.

▶ 8 Learn how to say no politely

There are lots of ways to do it, and it doesn't have to be rude. People who say yes too much pile pressure on themselves and may end up letting themselves, and others, down. Saying no will reduce the total amount of things you have to do, the number of interactions with machines, and will allow you to concentrate more on what you like doing.

▶ 9 Study how language works and learn as much as you can

Look at the dictionary regularly. Listen to great speakers. Read interesting books. Soak up the way words work. It will improve your word power and increase confidence. The more effective direct conversations you have, the less you will have to use technology.

▶ 10 Use the mate, mum or grandmother test

Test drive what you have to say on someone who knows nothing about it and doesn't care about it. If they don't get it, then you are not talking clearly enough. Start again, improve what you have to say, and then choose the most appropriate method with which to say it.

05: COMMU-NICATING

200 emails a day

THE DEATH OF HUMAN INTERACTION?

We have so many methods of communication available to us that not only do we frequently choose the wrong one, but also we often select a method before we have worked out whether that's the best way or not. Talking direct is always best, so everything else is a poor cousin, unless the message warrants nothing more than a quick exchange of information. It is important to remind ourselves that no particular technology is bad as such. It is our lack of thought in using the right type for the job that can be. Documents can remove character from a line of argument. Slides, if not well chosen, can detract from a presentation, and text and email can be the death of conversation.

It is easy to make email the whipping boy here, so let's put it in context. It's about volume and suitability. Those who receive too many emails put in deflective mechanisms. When John Freeman, author of *The Tyranny of Email,* started receiving more than 200 emails a day he knew things needed to change. Do the calculation and you will see that even if he spent just one minute reading and answering each one, it would take him 3 hours and 20 minutes a day. This is a clear case for revolution and reclaiming his life back. This sort of volume is probably more common in a work context, but the principle still applies.

Suitability is the second point. Nobody should send an email just because they can. As Stephen Fry said, 'THE EMAIL OF THE SPECIES IS DEADLIER THAN THE MAIL.' The medium of email is deficient in relation to talking in person. Entire companies have lost their soul by using the medium and nothing else. In training, I ask who has the most emails in their inbox. Five years ago the worst offenders had two or three thousand. The figure has steadily grown to the point where 10,000 is not uncommon. This year the record was smashed by someone who showed me their inbox total of 38,000.

Those who operate in this way are certainly EDICTED, and are probably technoholics. They are missing out on thousands of opportunities to engage properly with friends and colleagues, and may even be losing the art of doing so. Staff become ostracized from each other as a result. In one company I trained that had 250 staff, such a problem was identified and a new rule was introduced banning internal email on a Friday. Anyone sending one would pay £1 into the charity fund. One employee ended up paying £100. He simply couldn't stop.

Human interaction lies at the heart of a fulfilled life, and should be cherished and practised. If the volume of technological interaction has reached silly levels, we need to get our lives back. If our ability to choose the right medium for the message is often found wanting, then we need to think more carefully before communicating with the wrong method. Both are infinitely achievable if we pause and take stock. There are 50 or so resolutions in this chapter to help.

looking some-
one dead in the
eye

ROOM TO HIDE

One of the questions in the revolution test was: *Have you ever used technology to avoid talking face to face?* We know that technology has played a part in what many believe to be a decline in social skills, although it has helped in many others. Some people actively hide behind technology, and this is where the trouble starts. Before email was invented, the old office trick was to take a document, add an instruction to it, and put it on someone else's desk. Suddenly, it's not your problem. People do this with email all the time.

Hiding behind technology covers a broad spectrum, but falls into three main areas: indecisiveness, ego and cowardice. Let's start at the harmless, but nonetheless annoying end. If you want to fix something fast on email, then a specific proposal that requires a decisive answer is vital. So, 'SHALL WE MEET UP SOMETIME?' is hopeless, and will simply start a tedious trail of permutations whereas 'SHALL WE MEET? I CAN DO THURSDAY AT 2 OR 4, OR FRIDAY AT 10. CHOOSE ONE.' demands a clear and decisive response. One can immediately see that the onus is on the person starting the conversation to frame it correctly. This form of hiding is irritating and a waste of time, but not the end of the world.

Next up is ego. Studies by manufacturers such as Palm and BlackBerry have shown that the primary benefit of the hand-held device is to boost the self-esteem of the owner. This is fine for educational purposes to help classroom learning, but in the wrong hands these devices become weapons of status. They verify that the owner is busy and important, and many users end up genuinely believing that the world will fall apart if they cannot be contacted even for a brief period. This leads to an unfortunate chain of behaviour that can include disturbing everyone else in public places, ignoring people in face-to-face conversations, blanking members of one's own family, and failing to pay attention in meetings. They also seem to distort perceptions of time. Few people would call a landline in the middle of the night unless there was an emergency, but many do on mobiles. Weekend business calls are becoming more frequent too. The rap sheet is longer but the point is made. So many executives now check their emails under the table in meetings that it has become known as the BlackBerry Prayer.

The sharp end is that the worse the news, the more likely it is to come via a computer or device. That's cowardice. As we know, looking someone dead in the eye and delivering bad news is not easy. So now lily-livered management the world over has plenty of choice with regard to the medium they wish to hide behind. The all-staff email can cover a multitude of nasties, without the writer ever having to face the people affected. Those who are paid well to make tough decisions should examine the manner in which they communicate with their people.

READ IT, THEN ACTION, FILE OR BIN IT.

DON'T ASSUME YOU CAN HANDLE IT ALL AT ONCE

In days gone by, if you had a couple of letters in the post, or a few phone calls, then you could deal with them. But how many messages can someone deal with in a day? There must be a numerical limit. Wouldn't it be good if an email inbox would only take a maximum of, say, ten messages a day? The rest would wait in a queue until the recipient was ready for the next in line. The moral for individuals in the modern world is: don't assume you can handle it all at once. You might be able to, but if not, then it's your job to make some changes to how the communication is flowing in your life.

The power of sequence is important here. Doing one thing at a time means that it is easier to cope with large volumes. Never touch an email more than once. Read it, then action, file or bin it. In *Getting Things Done*, David Allen recommends turning your in-box upside down and working on the principle of First In First Out (FIFO), not LIFO as many people do. You need a system to cope with the huge volumes. Those who approach their in-box randomly will usually be swamped. A colleague used to sweep all her paperwork into the bin each night on the grounds that if it were that important, someone else would have it. Those who have had a technological meltdown and lost everything usually experience a short moment of panic, followed by extreme elation and relief. There are rarely repercussions either. Most colleagues and friends don't notice. If this is true, then deleting everything every few months could be a very good strategy for coping with communication indeed.

THINK BEFORE YOU DIVE IN

THE RIGHT MEDIUM FOR THE RIGHT MESSAGE

To communicate well with technology, we need the right medium for the right message, and a clear understanding of the suitability of each for the task. Intersecting monologues won't do. Almost everything we do involves the need for effective communication, and yet often we really aren't very good at it. Methods of communicating are always changing. The rank order of possible communication methods, based on the likelihood of you being correctly understood, is talking face to face, a telephone conversation, then an email, text or instant message. With regard to effectiveness, the first beats all the rest by a hundred to one.

Some people seem to think that once they have fired off a message they can wash their hands of something. Nothing could be further from the truth. Among the possible reasons for this are:

▶ The message never arrives.

▶ The intended recipient never reads it.

▶ They are not there.

▶ They are too busy or disorganized.

▶ They read it, but they don't agree.

▶ They want to discuss it with you before doing anything.

So as a high-quality communication method, email and other messaging methods leave much to be desired. But there are other more subtle possibilities lurking too:

▶ Most people don't check their messages before sending, so that any errors can make them look sloppy, or impede what they are trying to say.

▶ Your original message or reply is often forwarded to someone you don't know.

▶ The presentation format is often in the hands of the receiver, not the sender.

▶ People you don't know about are sometimes blind copied on the original for political purposes that you know nothing about.

The sort of chaos that can ensue from these possibilities shouldn't really require any further elaboration. Suffice to say that any communication method that has these pitfalls needs to be treated with extreme caution. Some careful thought will always lead to a better result. So think before you dive in, and try talking to people – it is much more charming, and almost always more effective.

a dying art

HOW TO WRITE CONCISELY AND CLEARLY

If you can't talk, then you are going to have to write. This is something of a dying art, so you need to put a decent argument together and get it down before you start using any form of technology to convey it. Don't just start writing. Draw up a shape and prepare a draft first. This applies to emails too. Though it may initially seem a lengthy process for a simple email, with practice it will become instinctive. Try following these steps.

▶ Work out what you are trying to achieve

Start by asking whether you need to write this down at all. Should this actually be a phone call or a meeting instead? What do you want to happen when someone reads this? Write down the objectives of the exercise, and, against each one, write the desired action of the reader.

▶ Research your intended audience

Who is going to read this? Who might read it that you haven't been told about? Are the readers similar in style and experience? What do they have in common? Are there unifying factors you can draw on which will make the readers more likely to agree? What attention span do they have? What frame of reference will best allow you to make your points? Are the readers potentially willing, reluctant or hostile? If they are hostile, what might persuade them?

▶ **Research your subject**
Do you really know what you are writing about? Could you know a bit more? If so, go looking for information. Are there sources that might help? Use positive arguments. Disproving the opposite case does not make yours. Using negative arguments makes you sound negative. Amass killer facts. They should be true, from a reliable source, and fundamental to your case. If you make assertions, consider whether you would believe them if you were the reader, particularly if they tend to the sceptical. If you are dubious, find the evidence to support your points, or change the phrasing to make them more opinion based.

▶ **Choose an accurate and engaging title**
If you are starting a debate, write a provocative question. If you are trying to persuade them of a particular stance, state it. If your purpose is information, keep it factual and free of loaded words. If you want to generate emotion, put it in before the subject matter. Play around with the chosen title and constantly check it against the relevance of the content as it develops. There is nothing wrong with changing the title once you have written the content, as long as it is an improvement.

▶ **Create a decent structure**
The better the plan, the better the writing. Try writing the complete argument in no more than a dozen points on one page. It needs an introduction, a beginning, a middle and an end. The introduction explains what

this is for. The beginning opens the case with relevant information. It should include a 'grabber' – something which makes the reader pay attention. This could be a quote, a fact, a comparison, something they don't know, or a controversial statement. This is followed by your main themes – preferably no more than three. The main themes may need supporting material or evidence. This is fine, but consider whether it interrupts the argument, or is essential. If it gets in the way, put it in an appendix or an attachment. The end needs to reprise the main themes, have a strong recommendation, and make it clear what you want the reader to do.

▶ Now add a bit of colour

Always ask yourself: have I heard this before? If you have, think of something more original. Imagine you are having a conversation with someone. How would you phrase it to them out loud? Don't make the language flowery. Add colour with relevant, distinctive words. When you have written something, read it out loud. If it sounds ludicrous, change it. Keep it simple and don't get weird, unless you are providing a counterpoint to the simplicity of your own proposal, or want to shock to make a point. Review the whole thing for colour, pace, rhythm and emotion.

▶ Avoid cliché and bullshit

As with the talking advice in the last chapter, imagine you were reading it to your mum or a mate in the pub. Do you feel stupid? If yes, change it. Would they understand it? If no, change it. Use the spotting waffle guide to check for false arguments, circular arguments, repetition, incorrect conclusions ('synapse jumps'), spurious sources, irrelevance and weak points. Most cases add weak points at the end that undermine the strong, valid points at the beginning. It is better to have two or three strong ones than a shopping list of increasingly weak ones. For an amusing diversion featuring some of the worst examples of business bullshit, have a look in the appendix.

▶ Edit several times

Shorter is always better. Be relevant, rational and ruthless. Remember the difference between a conclusion and a summary. A conclusion is when you have argued a point and concluded. You can do this on several different points to build your case. When you have made them all, the summary draws together everything you have said. It therefore repeats the conclusions, and repeats any recommendations. It is essential that the summary does not make any new points at all. If it does, your structure is wrong.

▶ Dealing with writer's block

Most people talk faster and more fluently than they write and can hear themselves a few sentences ahead of the writing on the page. So when you do not know what to write, try saying out loud what you want to say. Then write it down. In extreme cases, record yourself and listen back to it. Remember, writer's block is temporary. If nothing is happening, break the pattern by doing something physical, or different. If that doesn't work, do the most boring thing in your in-tray and come back to it later.

▶ **Check with someone else**

Give it the overnight test if you can. You will always spot something in the morning that you didn't the day before, and you will probably have had another idea. This is your own quality control test. Edit it ruthlessly, then show it to someone else. If they don't understand your main points, express them differently. Even when you are happy with the whole thing, get someone else to proof read it. Your eye may read misspelled words as correct because you wrote them yourself.

Once you have done all of this you are ready to transmit via whichever technology you have decided is the most appropriate.

Make it work for you

TECHNOLOGY MORALS

Communication is complex, so don't just dive in to the nearest technology and start transmitting. Think hard about how you want your life to be affected by it. Make it work for you. Tame it. Don't be a slave to it. There are many morals to be learned from the way in which much modern technology is used. Here are the most important.

- ▶ *Think hard about how you want it to affect your life.* Don't just accept that everyone else is doing it, so you have to too. Pause to decide what technology you would like to have as part of your armoury.

- ▶ *Make it work for you, not the other way round.* You don't have to use any technology you don't want to. Consider your ideal set up, and reorganize things to match that if possible.

- ▶ *Only turn it on for specific times each day.* Map out a regime, and stick to it. You will liberate free time to do lots of other interesting things.

- ▶ *Turn it off when you need to get things done.* Constant distraction simply leads to frustration and half-baked ideas and projects.

- ▶ *Don't assume you can handle it all at once.* You probably can't, so be honest with yourself and make changes.

- ▶ *Don't hide behind it.* Doing so suppresses your true nature and is usually unpleasant for everyone else.

- ▶ *Don't let it lure you into doing inappropriate things.* Illegal activities are clearly out, but also beware drunk and angry communication – no one wins.

- ▶ *Do not let it come between you and those you care about.* If you spend more time with a machine than your partner or family, then something is wrong.

- ▶ *Do not confuse what it can do with tasks that require humanity.* Always go for the human option before you use a machine to do the job.

- ▶ *Do not assume that any job is done by the technology alone.* It's not done until someone has done it. That's either you or the person you have asked, but bear in mind that they might not even agree, let alone do what you ask.

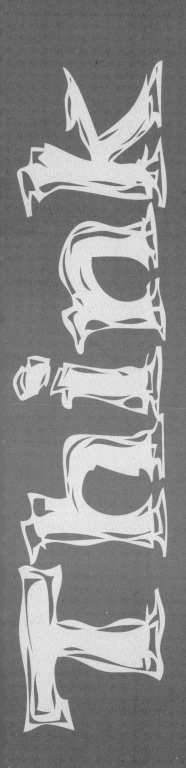

A NEW COMMUNICATION ETIQUETTE

You should aim to meet, call, write, and in that order. If you can't meet, then call. If you can't call, then write carefully. Here are some resolutions for each stage.

▶ **Meet**

- *When you secure a meeting, get organized straightaway.* If you leave it to the last minute you won't communicate as well as you want to.

- *Research everything thoroughly.* People who don't know their stuff get rumbled quickly and are less persuasive.

- *Give them a list of ways to fix things.* Alternatives provide lots to talk about and plenty of ways forward.

- *Include things that you could do, even if you have never done them.* That's how you get to do new things.

- *Ask what is on their mind at the moment.* Let people identify their own needs.

- *Listen more than you talk.* Otherwise you won't learn anything.

- *Be more positive than everyone else in every meeting.* Who wants to have a meeting with a boring, negative person?

- *Never be late.* It's rude, and suggests that you don't care. Just because you have a mobile that enables you to provide last minute updates does not mean you should use it as a smokescreen for poor manners or lack of promptness.

- *Be spontaneous and act naturally.* Pretending to be someone you're not never works.

▶ Call

▶ *Prepare what you are going to say.* What about x? Or y? Very well, what about z?

▶ *Don't use jargon.* If you waffle, they won't get it.

▶ *Tell it like it is.* Don't sugar coat things. If it's bad, be honest.

▶ *Offer to solve their issue quickly.* No one wants to wait if they need something.

▶ *Be natural and human.* Don't go into business mode. Be your normal self.

▶ *Don't call it 'cold calling'.* You might think it's cold, but the other person might not.

▶ Write

We have already covered a lot about how to write appropriately, but in summary:

▶ Only write if you cannot meet or call.

▶ Write concisely and clearly.

▶ Check it before you send it.

▶ If it's complicated or emotional, pause before sending.

Communication can be very complicated, so don't just dive in with any old words and any random technology. Think hard about what you want to say and how best to say it. Choose the most appropriate medium for the task, and pause to check everything before you send it. This will make everything you generate more thoughtful, and help you to communicate better all round.

TAMING TECHNOLOGY TIPS FOR COMMUNICATING

TAMING TECHNOLOGY TIPS FOR COMMUNICATING

▶ **1 Make human interaction your first port of call every time**

Always ask yourself: which is better for this job, a person or a machine? If it's the latter, then try to get some humanity into the process. Add some character, and make it clearly recognizable as your style.

▶ **2 Don't hide behind technology**

If it's tough and nasty, don't be a coward. People can take tough news, particularly if it is delivered with integrity and a little care. Do not use technology as a shield to avoid the issue.

▶ **3 Don't assume you can handle it all at once**

You might be able to, but you might not. Those who claim they can are often the ones who can't. There's no disgrace in saying enough is enough. In fact, it may even be a good thing to do.

▶ **4 Choose the right medium for the message**

Think carefully about the point of what you wish to communicate. Would this be better as a chat in person? Or a phone call? Is this email, text, or instant message really necessary? Does this need to be said at all? Take the extra moment to think about it.

▶ **5 Practise concise and clear writing**
Once you have decided that it needs to be written down, think carefully about the shape of it. Plan and draft before launching into something that isn't well thought-through.

▶ **6 Review your technology and make some changes**
Cast your eye over all your technology. Is it what you want? Does it work for you? Does it allow you to communicate effectively? If not, make changes that fall more in your favour.

▶ **7 Consider the new communication etiquette**
Meet if you can. Call if you can't meet. Write if you can't call, but think carefully about the content and method. In any event, try to increase the proportion of interactions that are personal rather than remote.

▶ **8 Develop ways to cope with techno-rage**
If technology drives you nuts from time to time, try to identify why and what the main culprits are. Then invent ways to reduce stress by using that technology less, under less time pressure, or by understanding better how it works.

▶ **9 If an email, text, or instant message conversation is getting out of hand, stop**
Conversations that escalate into ping-pong slanging matches are unhealthy. Recognize as soon as possible

when this is on the verge of happening. Resist the temptation to get involved in tit-for-tat exchanges. Break the pattern by picking up the phone, or turning up in person.

▶ **10 Don't dash to the nearest technology as a matter of course**
Don't make technology a default option. Work out what you wish to do first, or what you think. Then decide how best to communicate that to someone else. Don't start firing off instructions or messages until you know why.

06: DOING

Don't let technology do this to you

HUMAN BEINGS AND HUMAN DOINGS

We are human beings, but just being isn't enough for us. We need to be human doings. It is only by getting stuff done that we feel any sense of purpose and satisfaction. Every day we can choose from hundreds of activities, so the question is whether we are doing the right blend of things to make us happy. In relation to our technology, the answer will often be no. If we do too much of the wrong stuff, then that often means we do not have enough time left for the right stuff. We need to be able to distinguish between which parts of our relationship with our machines matter, and which don't, so we can make better use of our time.

There are two main types of task – quantitative and qualitative. If it's quantitative, the job just needs doing, full stop. You can tick the jobs off numerically and view them as a quantity of things to be done. If it's qualitative, the job needs doing well. The quality can be demonstrably better than if it is done badly, and excellence should be your goal. There is a massive difference between the two. It is crucial to take the time to distinguish between the two when you are sitting at a computer or about to make a phone call. If it is quantitative, then it should be fast and functional. If qualitative, take your time, plan carefully, and do it properly. These principles apply to any medium, and are particularly important if you are jumping between tasks and different technology (but try not to do this). Plan your tasks and spend an appropriate run of time using one method. Then pause, take a break, and move on to the next.

So what does *not* work is if you simply sit in front of your computer or pick up your device and start typing. There will be no shape to your efforts. Instead, take any of your lists and separate the tasks by those that simply need doing and those that need to be done well. It is hard to generalize about this because it depends somewhat on your job and lifestyle, but the percentage of functional tasks you typically have to do is likely to be high. Everything of this nature should be done as fast and clinically as possible, and with no delay. Many people like to do these tasks first, so as to allow proper thinking time for the quality jobs.

Should you choose not to take this approach, then it is highly likely that you will have a bar code day. This is a concept introduced by Matt Kingdon in a book called *Sticky Wisdom.* Picture a bar code. It is made up of many tiny slivers. Now imagine that this is a snapshot of your working day, or any unit of time, even an hour at your desk. If you blindly sit at a computer or device and try to cover as many tasks as possible, but in no particular order, you will be bar coding, or, put passively, you will be bar coded by technology. In other words, bittiness and interruptions will force you to work in smaller and smaller chunks, to the detriment of any proper attention on one important thing. Don't let technology do this to you.

Be smart and get on with it.

CONFRONTING THE DEVIL

I once had a conversation with the owner of a company that had grown over ten years to have around 600 staff. He was concerned that the original feel of a start-up had been lost underneath too much process and administration. It felt like the company was slowing down, and the suspicion was that too many people were doing a lot of things that they didn't need to do. Although the message from the top had always been to be smart and get on with it, the feeling was that this attitude was on the wane. The way he expressed it was that people 'NEEDED CONVINCING THAT THE DEVIL IS THERE'.

This is a crucial breakthrough moment in being able to do things effectively with technology, because many EDICTED people do not believe that there is a problem in the first place. Let's acknowledge that there must be some people who are so well organized, and have such a well-balanced relationship with their technology, that they can handle everything that's thrown at them everyday with comfort. For all the rest of us, we need a revolution, and when it comes to doing, we need to take three steps.

1 Admit that the devil is there.

2 Confront the issue.

3 Adopt a new attitude and approach.

So, step one is confession time. 'MY NAME IS KEVIN DUNCAN AND I'M A TECHNOHOLIC. I ADMIT THAT ON CERTAIN DAYS, IN FACT MORE OFTEN THAN I WOULD LIKE, I CANNOT GET EVERYTHING DONE THAT I WOULD LIKE BECAUSE I KEEP GETTING INTERRUPTED BY ONE FORM OF TECHNOLOGY OR ANOTHER. SOMETIMES IT'S

THE PHONE, SOMETIMES IT'S EMAIL, AND OTHER TIMES I JUST GET DISTRACTED POTTERING ABOUT ON SOCIAL MEDIA.' Try this exercise yourself.

Step two is confronting the technology devil. An endless bombardment of inbound material simply cannot be endured forever by one individual. Someone is going to crack, and it is more likely to be a human than a machine. And something important is going to slip through the net. The machismo language of the online world doesn't help. We are apparently supposed to be 'always on', available 24/7/365, or, according to one device manufacturer, capable of doing the impossible everyday. It's unsustainable and unrealistic. Those who have a balanced view are able to see it for what it is, and laugh at it by calling EDICTED people CRACKBERRIES because they have an addictive immediate response to every incoming message.

Step three is adopting a new attitude and approach, and that means setting up a new system. Your ability to get things done, whether involving a machine or otherwise, will improve dramatically if you make some simple changes. Start by setting aside thinking time. Use a system or draw up an organized list for tasks. Separate them by quantitative or qualitative. Divide them by those that require technology and those that don't. Map out your time to reflect the balance. Then, either do all the non-technology tasks before going near a machine, or do all the technological tasks first and then walk away from your computer or device to do the rest. This will lead to a more productive result in both online and ordinary activities, will prevent you from having a bar code day, and will liberate better quality time for the jobs that warrant it.

SOMEWHERE IN THE MID-
DLE SOMEWHERE IN THE
MIDDLE SOMEWHERE IN
THE MIDDLE SOMEWHERE
IN THE MIDDLE SOME-
WHERE IN THE MIDDLE
SOMEWHERE IN THE MID-
DLE SOMEWHERE IN THE
MIDDLE SOMEWHERE IN
THE MIDDLE SOMEWHERE
IN THE MIDDLE SOME-
WHERE IN THE MIDDLE
SOMEWHERE IN THE MID-

PROGRESS NOT PERFECTION

Is your very best perfect? I doubt it. Mine never is. This is the dilemma with striving for perfection. There is nothing wrong with it as a life philosophy, but it simply cannot be applied to everything that needs to be done. The more frenetic we are, the more we have to make do with 'good enough will do', and get on with it. Never was this approach more appropriate than online. The debate rages on two levels. The first (high level) theme is that a huge amount of unedited stuff appears on the internet. Fans say this is great because it generates a superb melting pot in which ideas can be exchanged and co-created. Detractors dismiss it as the random rantings of amateurs that are not subject to any editing. At the second (micro) level billions of messages are exchanged everyday that contain typographical and grammatical errors. Objectors feel this is a case of standards slipping, and fans say so what? We are communicating, so let us get on with it, and stay away if you don't want to join in.

As with so many things in life, perhaps somewhere in the middle is the place to be. If your online reputation is important, then make sure that the quality of what you produce is of a standard that befits what you are trying to achieve. Tasks can be undone by poor execution, so there is a link between quality and successfully getting things done. On the other hand, this can be pushed too far. Striving for high quality is great, but being a perfectionist can be a tremendous burden, particularly when the technology is there to make your life easier, not make it all more complicated. There are two fundamental problems with perfection:

1 Perfection may not exist.

2 Perfection may never quite arrive.

When it comes to technology, how can anyone ever prove that something is perfect? Perfection is a qualitative notion, and is therefore unhelpful for the purposes of getting things done. People often fall into the trap of making an ill-judged connection between success and perfection. And yet nothing could be more imperfect than those people and projects that eventually achieve success. Their road to achievement is usually littered with false starts and mishaps. Successful products are preceded by many prototypes, many of them defective. The more mistakes you make, the more you learn, and this is especially true of working with technology. It is changing all the time, and there is always so much more to discover, so it is crucial that you make mistakes in order to make progress. Striving to do something better is admirable, but do bear in mind that perfection may be a mythical construct than can never be achieved. As the old saying goes, if you want to succeed, double your failure rate. Or, in the words of Paul Arden, author of *Whatever You Think, Think The Opposite:* 'INSTEAD OF WAITING FOR PERFECTION, RUN WITH WHAT YOU'VE GOT, AND FIX IT AS YOU GO.'

plan
more
and
flap
less

ACTION NOT ACTIVITY

If you like a mantra to help you do things more effectively, try these:

▶ Efficiency is a sophisticated form of laziness.

▶ Don't confuse movement with progress.

▶ Take on less, do more.

▶ Action not activity.

▶ Outcome not output.

Efficiency is a sophisticated form of laziness means that the better organized you are, the easier it is to generate free time. You can then use that time for whatever you want. People who are constantly in essay crisis mode lack control and do not usually achieve what they want. So when it comes to your technology, plan more and flap less. Random events and creative thoughts are great, but you will have more of them if you create the right conditions.

Don't confuse movement with progress means don't get distracted by lots of movement that generates no forward motion. All talk and no action doesn't get you anywhere. How many times have you observed in life that a lot appears to be happening, but in fact, nothing much really is? This is what the Italians call the English Disease: rushing around creating the impression that things are happening, but with no real tangible results. You can sit at a computer all day and not really get anywhere, although your boss might think you are working hard. Do not generate a smokescreen of activity to disguise the fact that important things are not being done.

Take on less, do more means paying careful attention to the nature and scope of what you are doing at any given time. Technology can be deceptive. It can hide a mountain of tasks because there is no visible untidy desk. Who knows how many new tasks have arrived in your email inbox at the same time as your acceptance of a busy new project? Screen all your machines for new work before accepting more, and consider saying no politely.

Action not activity means getting to the decisive point of action rather than diving into a blizzard of activity. They are not the same thing. What's so clever about being busy? Any fool can appear to be permanently busy, or truly be busy, particularly if they hide behind technology. If you want something to happen, concentrate on the action, not on activity that makes it look as though action is occurring.

Outcome not output is a different way of expressing the same thing. A lot of output makes it look as though much is happening, when frequently it isn't. If you achieve something excellent, who cares how you got there? If you have a great idea, who cares whether it happened in a flash, or over two weeks, or several years? Generating huge amounts of stuff (output) may have nothing to do with the end result (outcome). So before you spend days at a computer producing endless documents and spreadsheets, or firing off email after email, work out your desired outcome and work backwards from there.

like a mosquito
in a
nudist camp

MULTITASKING AND RAPID SEQUENTIAL TASKING

A couple of years ago I had just put the bread in the toaster when my partner asked me a question about holiday plans. 'I'LL THINK ABOUT THAT IN A MINUTE,' I replied, 'I'M JUST MAKING TOAST.' I wasn't joking. It's classic male stuff – we can only do one thing at once apparently. The debate rages on to the point of cliché – women can multitask and men can't. As a male myself I am happy to concede the point: I am not good at doing several things at the same time, and technology has now made this problem a whole lot worse. When confronted by many panels on a computer, I am like a mosquito in a nudist camp. I know what I want to do but I don't know where to start. So what can men do if this is truly the case? How can they be any good at getting lots of stuff done all at once?

The secret lies in RAPID SEQUENTIAL TASKING. Just because men can't do lots at the same time does not mean that they can't do lots in a sequence, and fast. Tackling the problem of the 'DON'T TALK TO ME, I'M MAKING TOAST' syndrome involves doing things fast, but one after the other. If you have ten emails to send, then do them one at a time in one half-hour session, then close the email function. If you have three presentations to write, then have a PowerPoint session, then close the software. If you need to do some accounting work, spend an hour on spreadsheets and then close it down. If you need to make six phone calls, then move away from your desk and do those one at a time. Changing media like this is good for variety, and may even force you to take a short break in between each blast.

This sequential approach is far more effective than having everything on the go at once. It also has one distinct advantage over multitasking. As we saw in Part I, studies show that the most persistent multitaskers perform badly in a variety of tasks. They don't focus as well, they are more easily distractible, and they are weaker at shifting from one thing to another. In fact, they are worse at it than people who do not usually multitask. There is a strong suspicion that in the case of much multitasking, all the tasks may well have been started, but they may not have been finished. This is a crucial point. Although the beginning of any task is clearly vital, it isn't over until it's over. The beauty of Rapid Sequential Tasking is that you don't move on to the next task until you have finished the last one, making it much easier to cope with whatever technology throws at you next.

Do less and get more done

ANTI LISTS

An anti list is a list of what you are *not* going to do. This is a crucial aid to taming technology and establishing what you *are* going to do. There are various ways in which writing this list can really help. It establishes:

1 What you will *never* do with your technology.

2 What you don't *want* to do with your technology.

3 What you won't do *today* with your technology.

4 What action really *will* help achieve the task (with or without technology).

All are tremendously helpful to know, and could be equally valid, depending on the nature of the job. Firstly, it is crucial to know what you will never do with your technology. Whatever these things are, they are vital components of your standards, principles, personal character, and, vitally, your relationship with your machines. If you have never done the exercise before, take a sheet of paper (or tap into your laptop or device), 'I WILL NEVER...' and fill it in the rest of the sentence. There may be several pledges. They can be relatively small, such as 'I WILL NEVER BE LATE FOR WORK AGAIN', or large principles such as 'I WILL NEVER WORK FOR COMPANY OR PERSON X AGAIN.' It's a very therapeutic process.

Secondly, you will clarify what you do and don't want to do. We all have to do things that we don't really want to, and obviously some are much worse than others. Identifying what you don't *want* to do with your technology, versus what you will *never* do, is a very helpful comparison.

The third point, what you won't do with your technology today, is a temporal one, and one of priority. Prevaricators who make a life's work out of putting everything off require significant help here. Tasks do not improve in quality if they are delayed. The value of establishing what you won't do today is so that you can do more important things first, not so that you never do them. Also, today is just one unit of time to describe when the task will be done. It could equally apply to the next five minutes, the next hour, this morning, tomorrow, this week, this month, or this year. Don't become a victim of time. You must be acutely aware that the longer the unit of time, the less likely it is that the task will be done.

Finally, consider what action really will help to achieve a task. It may not require any technology at all. Draw up some principles and attitudes to your machines, and try hard to stick to them. Be stubborn. Do less and get more done.

DOING NOTHING AS A FIRST OPTION

WHEN DOING NOTHING IS BEST

As far as we can tell, it is generally better to do something rather than nothing. Presumably that's because action gives us a sense of purpose, and inaction suggests idleness. But here's a contrary view – sometimes doing nothing is the smartest move. This is a view held by Ofer Azer, a lecturer at the School of Management at Ben-Gurion University of the Negev in Israel. He argues that, when people are under pressure, the urge to take action is powerful. Goalkeepers who let in penalties feel better if they have at least moved. Traders losing money on shares feel better if they sell, even if they make a loss. And politicians are always tempted to 'do something' when the economy is doing poorly. When it comes to technology, we should certainly choose the 'do nothing' route more often.

One man who has an interesting take on this whole area is Ricardo Semler, a Brazilian who runs a massive set of companies, and insists on working in an unconventional way. He likes to question everything, and in his book *The Seven-Day Weekend* he asks, among other things:

▶ Why are we able to answer emails on Sundays, but unable to go to the movies on Monday afternoons?

▶ Why do we think the opposite of work is leisure, when in fact it is idleness?

There are two levels here. One is that doing nothing can be a vital part of work, and vice versa, as we saw with the Bleisure Time phenomenon in Part I.

Even when you have a vast amount to do, it pays to pause and reflect. If work can intrude on your leisure time, then leisure should be allowed to play a part in your work. If introduced appropriately, it should improve your relationship with your technology. The second level is the balance between work and idleness. Tom Hodgkinson edits a magazine called *The Idler*, so he should know all about it. As far as he is concerned, society today extols the virtues of efficiency and frowns upon laziness, but as Oscar Wilde once said, 'DOING NOTHING IS HARD WORK'. As modern life grows ever more demanding, we may well feel the odds stacking against us, so we need an antidote to the work-obsessed culture that puts so many obstacles between us and our dreams.

There are benefits to doing nothing. For example, lying in bed half awake – what sleep researchers call the hypnagogic dream state – is positively beneficial to health and happiness, and can help prepare you mentally for the problems and tasks ahead. It is also the time when some of our best ideas come to us. The rational 'overmind' largely ignores the emotional or spiritual 'undermind', but this is where we build up strength to cope with life's struggles. Nobody knows why but sleep can solve many of our problems. Apparently insurmountable problems almost always look better in the morning. As John Steinbeck said: 'IT IS A COMMON EXPERIENCE THAT A PROBLEM DIFFICULT AT NIGHT IS RESOLVED IN THE MORNING AFTER THE COMMITTEE OF SLEEP HAS WORKED ON IT.'

Doing nothing is always worth considering as a first option. We often feel (or are made to feel?) guilty about taking time off, and we shouldn't. Americans now work an extra month a year compared with 30 years ago, averaging nine hours a day. Life is supposed to be getting easier, but we still elect to overwork. As such, those overdoing it in the workplace would do well to take a little time out, if only to check whether they lack balance between work and relaxation. Doing nothing with your machines must surely be worth consideration, if only for a short while each day or week, just to change the pattern.

Confront the doing part of your life in the context of technology. Distinguish between how you are going to get things done and the most appropriate method for doing it. Work out what you are *not* going to do, and make best use of your time to do the most important things.

Taming Technology
Tips for Doing

TAMING TECHNOLOGY TIPS FOR DOING

▶ **1 Admit that the devil is there, and confront it**
You may be perfectly happy with the way you do things with technology. But if it is all a bit overwhelming, it might be worth admitting there is an issue, and confronting it.

▶ **2 Adopt a new approach to doing things with technology**
Change the pattern. Plan your approach first. Group together tasks that require the same technology. Estimate how long it will take and try to set time limits. Move smoothly from one group of jobs to another, and take breaks in between.

▶ **3 Distinguish between quantitative and qualitative tasks**
Quantitative tasks can neither be done well nor badly – they just need to be done. Get these out of the way first and fast so that you can free up time to concentrate on the important, qualitative, tasks, whether you are using technology or not.

▶ **4 Go for progress, not perfection**
Don't spend hours at a computer making something absolutely perfect. Do your best of course, but remember that perfection is a matter of opinion. Someone else may well have a different view and ask

for changes anyway, so give it your best shot and fix it as you go along. You'll get more done and be less stressed.

▶ 5 Concentrate on action, not activity

Just because a lot is going on doesn't necessarily mean you are getting anywhere. Always ask yourself *why* you are using a certain machine and *what* you are trying to achieve. One precise action could be more effective than a day's worth of frenetic activity.

▶ 6 If you can't do multitasking, try Rapid Sequential Tasking

If you can do it all once, then go for it. If not, then don't try. Take one thing at a time, and don't start anything else until you have finished it. Then move on to the next.

▶ 7 Try writing an anti list

Lists are great, but if there is a lot on them, they can be daunting. Try writing a list of what you are *not* going to do. It is not a waste of time. It tells you a lot about your priorities with technology, and is great for morale.

▶ 8 Take on less, and do more

When a machine asks or tells you do something, consider saying no, politely of course. People who take on too much usually let someone down, and that someone may well be you. By taking on less, you will get more done of the things that really matter to you.

▶ 9 Consider doing nothing

Every now and again, take no action whatsoever. How many times has an issue been resolved after a trail of ten emails, before you even answered the first one? Let things go hang sometimes, and try to do nothing with your machines for a certain period every day, and every week.

▶ 10 Think about doing and being

Pause occasionally and consider the link between what you do and who you are. Actions say a lot about you, and with so many of these arriving via various forms of technology, they can quickly become your outward persona. Do you define who you are, or does your technology?

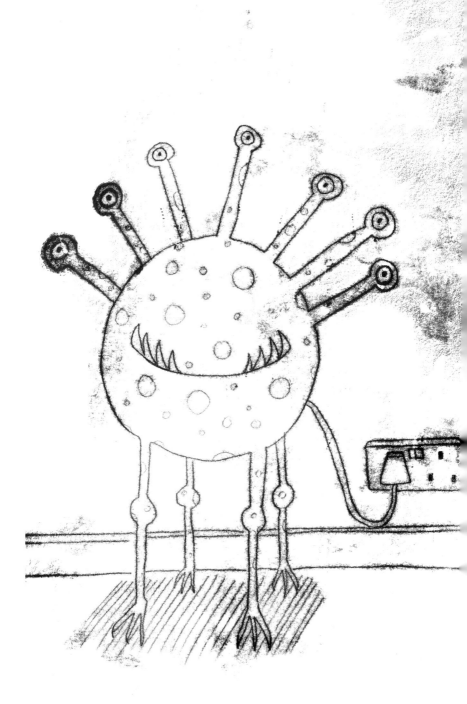

07: BEING

what defines **you?**

DEFINED BY YOUR CHARACTER OR YOUR TECHNOLOGY?

We have looked at thinking, talking, communicating and doing, and seen that a balance is needed between engaging in any of these activities with or without our machines. And so we come full circle to the very essence of our existence – being. As Anton Chekhov pointed out, 'ANY IDIOT CAN FACE A CRISIS. IT IS DAY-TO-DAY LIVING THAT WEARS YOU OUT.' So if you are completely obsessed with technology and could not live a day without it, we come to the nub of it all: what defines you? If the answer is you, then you have it right. If the answer is any kind of technology, then some kind of revolution may well be needed.

Specialists who make their living from technology as experts may well be exempt from this assertion. Software developers, social media gurus and other digital natives will of course be working with machines all day. They are, however, entitled to a life outside that work, and may find it therapeutic to have hobbies and outside interests that have nothing to do with technology. But for most of us, technology is an enabler that helps enormously if we use it judiciously, and a monster if we let it rule our lives. When it comes to being, we can all be better. We should define our own characters, not allow them to be defined by our possessions or technology. Having a better life starts with knowing what you are all about, and emanating that style.

Ask yourself what you are like, what you are good at, and what you wish to stand for. There are simple exercises that you can do to define your character. Ask yourself:

▶ If someone met you for the first time, how would they describe you?

▶ How would you describe yourself to someone you have never met?

▶ Are there differences between your work and outside personality?

▶ Is your inner self significantly different from your outward persona?

▶ Does your relationship with your technology affect your approach to life?

You can define your own style by asking yourself:

▶ Who or what is your favourite person or team?

▶ What qualities make them so good?

▶ How can those qualities inspire your approach to technology?

The stronger your sense of self-determination, the less likely you are to be dictated to by your machinery.

A RATIONAL ACTIVE LIFE

IN PURSUIT OF EUDEMONIA

EUDEMONIA is the happiness resulting from a rational active life – a concept first introduced by Aristotle. It's a life-affirming idea that deserves attention and discussion in our pursuit of a manifesto for a better life. Aristotle asserted that the value of moral action lies in its capacity to provide happiness. In other words, what we do defines who we are. Here the link between doing and being is completed – the way we behave has the capacity to dictate our happiness, or lack of it. So if we think carefully, and then communicate and act appropriately, we will be determining our quality of life directly.

A eudemon was believed to be a benevolent, in-dwelling spirit – your soul, if you like. Eudemonics is therefore the art or theory of happiness. Technology may or may not have any role to play in this. It is up to the individual to determine whether it could or should. One thing is certain though: technology cannot be the *only* thing that determines a person. The key lies in getting your attitude right so that you emanate a strong view. The machines can fall in line with that, not the other way round.

Aristotle also believed that it is the mark of an educated mind to be able to entertain a thought without accepting it. In the spirit of that, it is not necessary for you to agree with everything in this book, but if you grab a resolution or two and improve your relationship with your technology even a little bit, then the exercise will have been worth it.

YOU'RE
THE BOSS.
YOU'RE
THE BOSS.
YOU'RE
THE BOSS.
YOU'RE
THE BOSS.

GETTING YOUR ATTITUDE RIGHT

There are lots of things a person can do to get their attitude right and put some character into the way they operate. How you conduct yourself is crucial to the degree to which you can tame technology and get your life back. Consider these attitudinal resolutions.

▶ **You determine your own culture**
You're the boss. Review your relationship with your machines. Ring some changes. Tell colleagues, bosses and family that you are taking back control. Decide on a new regime that suits your style more and makes you feel better about yourself.

▶ **As far as possible, only do things that you like**
A pipe dream? Not necessarily. The apocryphal wise advice from an older lady to her daughter was to find something you love doing and find someone to pay you to do it. The more you like what you do, the less it feels like work. Design a blend of technology that gives you the same sensation.

▶ **When it comes to machines, do not distinguish between nice and nasty things to do**
Sometimes stuff just has to be done, so don't waste time agonizing over how dreary you feel it is going to be. Instead, look at your machines and work out how they can take more of the burden. That's what they are there for. Consider all the tasks that you find

tiresome, and work out how technology help. You might be surprised at how liberating this can be. For example, if you often forget to pay your bills on time, enter a monthly reminder once in your phone and let the machine do the rest, or set up direct debits so you never have to remember again. If you are dreading making a phone call, don't pre-judge the outcome. It may turn out to better than you thought.

▶ Remind yourself of all the positive things that technology can do

Viewed another way, try not to regard your machines with disdain. If you really hate any of them, then by all means get rid of them. But also imagine life without them and consider the alternative. This process could yield two sets of responses: confirming that you *can* do without something, and confirming that you *can't*. Both are valuable discoveries.

▶ Never use a piece of technology unless you know why you are using it

This sounds obvious and yet it may not be. Much of what we do is habitual or irrational. Why do I always have that cup of tea or coffee at that time? And why in that format? Why do I smoke? Or drink? Or check email repeatedly? Or eye my phone for messages all the time? A lot of these actions are less conscious than we think. By questioning every technological encounter, we can understand our relationship with our machines better and do something about it.

humility

+

honesty

+

humour

=

happiness

HUMILITY + HONESTY + HUMOUR = HAPPINESS

Three Hs – humility, honesty and humour – can contribute significantly to a fourth one: happiness. Being well adjusted helps you to cope with extremes, to laugh at mishaps and to shrug off adverse conditions. Humility, honesty and humour help massively in this struggle. They are your force field. As Tom Peters points out, the world will not come to an end if you are out of touch for 20 minutes. As a general rule, it pays to remember that it's just possible that you may not be vital to absolutely everything that goes on.

Humility, or the ability to be humble, may sound like a rather strange quality to recommend as a constituent part of technological success, but let me explain. Humility has a number of subtly different meanings – being conscious of one's failings, being unpretentious, and being deferential or servile. Being conscious of one's failings is crucial. So is being unpretentious. Being servile, however, is not desirable. Why is it helpful to be conscious of one's failings in relation to technology? So that you can use it to compensate for your weaknesses. Lack of pretension is a highly desirable quality too. The world these days is full of it – full of 'better than thou' people, frequently with the condescending language to match. Talk straight, and people will appreciate it hugely.

Honesty is important on four levels when it comes to taming technology.

- ▶ Being honest with yourself.
- ▶ Being honest about your approach to technology.
- ▶ Being honest when communicating through it.
- ▶ Being honest with others as an end result.

Try to analyse your technological relationship in a rational way. Try not to cloud it with inaccurate self-perception. Design an approach that suits you and that is realistic for your capability, set-up and available time. Retain your integrity when communicating with machines, and regularly realign the one with the other. The net result should be that there is no difference between dealing with you direct and dealing via technology.

Humour is crucial to wellbeing. Read any medical bulletin and it will tell you that a good laugh is good for your health. Besides which, it makes everything more fun. Who wants to hang around with a curmudgeon? The world has its fair quota of dull, worthy people, so why would you want to join in? You don't need to wear a comedy nose all day or practise stand-up routines in front of the mirror to entertain everyone. Just view the world with a lighter touch. Awkward situations can be diffused brilliantly with a smile and a humorous attitude.

Humour and common sense go hand in hand. As William James said: 'COMMON SENSE AND A SENSE OF HUMOUR ARE THE SAME THING, MOVING AT DIFFERENT SPEEDS. A SENSE OF HUMOUR IS JUST COMMON SENSE, DANCING.' Common sense is an absolutely vital ingredient in handling your technology. And as we know, it isn't all that common. A healthy sense of humour will allow you to view your interaction with machines with a wry smile. So when you have a problem with it – a technical issue, some kind of meltdown, or just a misunderstanding born out of poor communication – you will be able to see it for what it is: a temporary storm that can be resolved with a laugh and the right attitude.

The three Hs, leading to a fourth: humility, honesty and humour equal happiness. Take the issues seriously, but not yourself.

TAME TECHNOLOGY_
GET YOUR LIFE BACK_

DURATION, VARIATION AND VACATION

When it comes to being, there are certain things you can do to be relentlessly enthusiastic. Keep an eye on duration, variation and vacation. That means never doing one thing for too long, having plenty of variety in what you do, and going on holiday at suitable intervals. This will enable you to set your standards high, keep them there, and enact them everyday with a high degree of consistency.

It is a rare person that enjoys doing the same thing over and over again for a very long time. That could mean several hours at a computer on the same day. Or it could mean most days of the week for three months, or most weeks of the year for five years. The ratio doesn't matter, but the principle does. Eventually we all get bored. Consequently, it is very important that you never do one thing on a machine for too long. In the context of one working day, it is probably unhelpful for you to do one particular task for more than a couple of hours. To stay fresh, you should move on to something else unless it is one of those exceptional items that simply has to be churned through from time to time and really does take a long while. Even then, you may still need regular breaks from it, and breaking up any monotonous task is a healthy thing to do.

In any particular working week, you really do not want to be having the same technological encounter every day. You can keep it up for a while, but not for months. Keep reminding yourself that you are the person in charge. If your work is becoming repetitive, change it.

Benjamin Franklin decreed that 'THE DEFINITION OF INSANITY IS DOING THE SAME THING OVER AND OVER AGAIN AND EXPECTING DIFFERENT RESULTS.' If that's you, then you need to engineer a set-up that keeps you sane. To recap:

▶ Keep lots of variety in what you do to stay fresh.

▶ Change things if you don't find them interesting.

▶ Take regular breaks and a sensible amount of time off.

If you do this you will be able to approach all your dealings with technology energetically and enthusiastically. That's the vital importance of balance in your life. So to recap, you need to look carefully at the five areas and make some revolution resolutions:

THINKING. We don't do enough of it, even though it's completely free. Events overtake us. We need to rediscover the art of thinking clearly and use it to improve our quality of life.

TALKING. We do too much of it, often without having thought first. We talk too much rubbish, and not enough sense. It's time for a new, more considered approach that reflects what we feel more accurately and makes it easier for others to understand us.

COMMUNICATING. We have so many methods of communicating available to us that we frequently choose the wrong one. We need the right medium for the right message, and a clear understanding of the suitability of each for the task.

DOING. We do far too much of the wrong stuff, which often means we do not have enough time left for the right stuff. We need to be able to distinguish between what matters and what doesn't to make better use of our time.

BEING. We can all be better. We should define our own characters, not allow them to be defined by our possessions or technology. Having a better life starts with knowing what you are all about, and emanating that style.

It's time to tame technology and get your life back. As Bob Dylan said: 'A MAN IS A SUCCESS IF HE GETS UP IN THE MORNING AND GETS TO BED AT NIGHT, AND IN BETWEEN HE DOES WHAT HE WANTS TO.'

Taming Technology Tips for Being

TAMING TECHNOLOGY TIPS FOR BEING

▶ **1 Define your character and style**
Take the time to define and understand what you are all about. Only then will you be able to plan and enact your relationship with your technology.

▶ **2 Consider the idea of eudemonia**
It's a big word but it's not a complicated idea. Work out how the way you behave with your machines affects your happiness, and make changes based on what you conclude. Aim for a rational, active life, with the emphasis on the rational when it comes to technology.

▶ **3 As far as possible, only do technological things that you like**
Your work may force you to do some things that you don't particularly enjoy, but try to keep this to a minimum, and certainly do not replicate these tasks in your free time.

▶ **4 When it comes to machines, do not distinguish between nice and nasty things to do**
Some things just have to be done and there's no getting away from it. Review how technology can do these tasks for you and so relieve the burden. If it can't, then just get on with it.

▶ **5 Remind yourself of all the positive things that technology can do**

Imagine a world in which none of your technology existed. Use that as a reminder of how helpful a lot of it is.

▶ **6 Never use a piece of technology unless you know why you are using it**

More thought and less diving in. Many of us have irrational habits, and the way we use our technology may well be one of them. Don't take this for granted. Pause, reflect, and consider if there is another way.

▶ **7 Use humility, honesty and humour to create happiness**

The four Hs are a powerful blend. Recognize your failings and use technology to compensate for them. Be honest with yourself and your relationship with your devices. Have a laugh more often. You'll be happier.

▶ **8 Keep a careful eye on duration, variation and vacation**

Don't do anything for too long, keep things varied, and take breaks at suitable intervals.

▶ **9 Smaller chunks of technology mean greater clarity**

Don't be seduced by huge projects. Pose a problem, solve it with the most relevant machine, and then move on to the next (preferably small) thing.

▶ 10 Take the issues seriously, but not yourself

No further explanation should be needed. Things can be serious, but you don't have to be. It will be alright, honestly.

RESOURCES

REFERENCES

▶ Chapter 1: The Road to Ediction
Affluenza, Oliver James (Random House, 2007)

Cognitive Surplus, Clay Shirky (Allen Lane, 2010)

Enough, John Naish (Hodder & Stoughton, 2008)

Faster, James Gleick (Abacus, 1999)

National Phobics Society Survey (2009)

Socialnomics, Erik Qualman (John Wiley, 2009)

The Age Of Unreason, Charles Handy (Arrow, 1989)

The Paradox of Choice, Barry Schwartz (2004)

The Play Ethic, Pat Kane (Pan, 2004)

The Selfish Capitalist, Oliver James (Vermillion, 2008)

▶ Chapter 2: Do you need a Revolution?
American Journal of Psychiatry (2010)

Generation Me (Association for Psychological Science, 2010)

HotJobs Yahoo Survey (May 2009)

Information Anxiety, Richard Saul Wurman (1990)

In Search of the Obvious, Jack Trout (John Wiley, 2008)

Internet Innovation Alliance (2010)

Sunday Times (29 March 2009)

The Cost Of Multitasking (Ophir & Nass, Stanford University)

The Little Big Things, Tom Peters (Harper Collins, 2010)

The Shallows: What the internet is doing to our brains, Nicholas Carr (2010)

The Times (31 June 2009, 8 May 2010)

The Tyranny of Email, John Freeman (2009)

▶ Chapter 3: Thinking

Execution, Bossidy & Sharan (Crown Business, 2002)

Introducing Psychology, Nigel Benson (Icon, 1998)

Small Business Survival, Kevin Duncan (Hodder & Stoughton, 2010)

So What?, Kevin Duncan (Capstone, 2008)

The Cost Of Multitasking, Ophir & Nass (Stanford University, 2010)

The God Delusion, Richard Dawkins (Bantam Books, 2006)

The Tipping Point, Malcolm Gladwell (Little Brown, 2000)

▶ Chapter 4: Talking

High Impact Speeches, Richard Heller (Pearson, 2003)

Rework, Fried and Hansson (Vermillion, 2010)

Tick Achieve, Kevin Duncan (Capstone, 2008)

▶ **Chapter 5: Communicating**

Business Greatest Hits, Kevin Duncan (A&C Black, 2010)

Getting Things Done, David Allen (Piatkus, 2001)

Run Your Own Business, Kevin Duncan (Hodder & Stoughton, 2010)

The Paradox of Choice, Barry Schwartz (2004)

http://www.glencoe.com/sec/teachingtoday/educationupclose.phtml/14

▶ **Chapter 6: Doing**

Do Nothing, New York Times (9 March 2010)

How To Be Idle, Tom Hodgkinson (Penguin, 2004)

How To Get More Done, Fergus O'Connell (Pearson, 2008)

Marketing Greatest Hits, Kevin Duncan (A&C Black, 2010)

Sticky Wisdom, Matt Kingdon et al. (Capstone, 2002)

The Seven-Day Weekend, Ricardo Semler (Century, 2003)

Whatever You Think, Think The Opposite, Paul Arden (Penguin, 2006)

▶ **Chapter 7: Being**

Obliquity, John Kay (Profile Books, 2010)

Small Business Survival, Kevin Duncan (Hodder & Stoughton, 2010)

Start, Kevin Duncan (Capstone, 2008)

APPENDIX: CLICHÉ AND JARGON RED ALERT LIST

(Add your own over time)

24/7/365
Acid test
Added value
Any artificial -ize verb (optimize, diarize, prioritize)
Any word attached to -driven (e.g. customer-driven)
Anything to do with focus
At the end of the day
B2B or B2C
Ballpark figures
Benchmark
Best practice
Bottom line
By the end of play
Cold, hard facts
Core competencies
Cut through
Cutting or leading edge
Deliverables
DNA
Empowerment/enablement
Fast-track
Frameworks
Get into bed with
Get our ducks in a row
Give it traction
Go for the low hanging fruit
Going forward

Hit the ground running
Holistic
Innovative
Inputs (especially as a verb)
It's not rocket science
Key/key drivers
Leverage
Missions and tasks/mission critical
Multiple core objectives (there is only one core)
No-brainer
No stone unturned
Offline
Overviews and scenarios
Paradigm shift
Play a vital part
Proactive
Pushing the envelope
A raft of proposals
Right across the board
Ring fenced
Seamless
Silver bullet
Singing from the same hymn sheet
Stakeholder
State of the art
Step change
Synergy
Take a long hard look
Take this offline
Teeing up
There are no easy answers
There's no 'I' in team

Thinking outside the box
Touch base
Umbrella
Wash its own face
Win-win

For regular updates, bullshit books and apps, visit expertadviceonline.com

INDEX

24-hour news media   21

action and activity   168–9, 182
Affluenza (James)   24
alerts   71, 94–5
all at once, handling/doing
    53, 135–6, 147, 153, 171
Allen, David   135
Americans
    and consumerism (Fromm)
        24
    and texting   41
    and working hours   179
Aniston, Jennifer   38
anonymity and technology   56
anti lists   174–5, 182
anti-social behaviour   38
Apple
    iPhone (milestone)   15
    iPod (milestone)   13–14
    product sales   32
Arden, Paul   166
Aristotle   190
Asus Eee (milestone)   15
Attention Deficit Syndrome   47
attitude, getting it right
    192–3
audience research   141
Azer, Ofer   177

'bar code day'   160, 163
being   74, 201
    and doing   183
    as resolution   185–205
belief versus interests   76
'big picture'   10, 73
BlackBerry
    banning in meetings   60
    as milestone   14
    Tom Peters on users   57
BlackBerry Prayer   133
Bleisure Time   17–18, 21, 177
blind copying   139
body language   59, 120
Brachylogy   99–100
bragging rights   57
breaks, taking   172, 181, 199,
    200, 204
brevity and intelligence   99, 124
bullshit   143–4, 213

Call (New Etiquette)   150
call (phone), when to   120
Castle, Barbara   90–1
character, defining own   187–8,
    203
Chekov, Anton   187
choice, effects of more   27–8
circular arguments (waffle)   106

clichés (waffle)   107, 143–4
  alert list   211–13
Cognitive Surplus (Shirky)   10
common sense   196
communicating   72–3, 200
  as resolution   127–55
communication
  face to face   45, 120
  and language   103
  and straight talking   102
  use of media/methods   119–20, 129
concise, being   99, 112, 123, 141–5, 151, 154
conclusion
  incorrect (waffle)   106
  versus summary   144
consumption
  over-consumption   20–2
  technology as   18
Continuous Partial Attention (CPA)   38
convergence, effects of   28
conversations   37, 112
  and email   44–5, 129–30
  face-to-face   44, 120, 133
  getting out of hand   154–5
  and wafflers   112–13
cowardice and technology   133
Cowell, Simon   24
Crackberries   163
'cuddle' chemical   60
culture, determining own   192
Cyburbia   47

Decade of the Gadget   13
decisive, being   132
definition of self   187–8
Descartes, Rene   74
desocialization   45

devices *see* hand-held devices; mobile phones
devil, confronting the   162–3, 181
digital universe, size of   10
discontent and earnings   21
distraction   47–8
  and multitasking   54
Distraction Overload   48
doing   73, 200–1
  and being   183
  nothing   177–9, 183
  as resolution   157–83
  things you like   192, 203
dopamine and purchasing   20
driving
  and mobile phones   47–8
  and texting   42
duration and being   199–200, 204
Dylan, Bob   201

earnings (income/wages)
  average over time   24
  and discontent   21
  'enough'   25
Ediction   7–8
  and Affluenza   24–5
Ediction Test   62–5
editing out   100
editing writing   144, 145
efficiency and laziness   168, 178
ego
  and hand-held devices   132–3
  and saying 'no'   115
Einstein, Albert   119
email   44–5
  attention paid to   44–5
  as death of conversation   129–30
  'read, action, bin'   135
  and selling   59

Stephen Fry on   129
when to   120
see also messages
emotional needs   25
engagement
with people   60, 130
of self (brain)   25, 119
with technology   79, 107
Enough (Naish)   20
enough, developing sense of   20–1
etiquette
in meetings   59
New Communication Etiquette   150–1, 154
Eudemonia   190, 203
exabytes   10
explaining test   119
external waffle   105
eye, looking in the   56, 133
eye contact   37, 59, 120

Facebook   11, 47, 57
false arguments (waffle)   105–6
filtering out   100
Flip (milestone)   15
forwarding messages   139
Franklin, Benjamin   200
Freeman, John   44, 129
Freizeitstresse   50–1
'friends' on Facebook   57
Fromm, Erich
Fry, Stephen   129
functional tasks   160

gadgets
'decade of'   13
psychology of   38
Generation Me   44
Getting Things Done (Allen)   135
Ghandi (on speed)   25

Gladwell, Malcolm   82
Gleick, James   28
'God is in the details'   82
'grandmother' test   119, 125
Griffiths, Richard   38

hand-held devices
in business meetings   59
and self-esteem/ego   57, 132–3
Handy, Charles   7
happiness   22, 24, 204
and three Hs   195–7
Hemmingway, Ernest   112
hobbies and relationships   31
holidays   51, 199
honesty   195–6, 204
hours worked in lifetime   7
human interaction   60, 129–30, 153
humility   195, 204
humour   196–7, 204

idleness and work   178
Idler (Hodgkinson)   178
In Search of the Obvious (Trout)   59
inbox
and First In First Out   153
as 'slot machine'   44
income *see* earnings
inconsistency (waffle)   107
incorrect conclusions (waffle)   106
indecisiveness   132
Infobesity   21
Information Anxiety   50
information overload   11
inner self and persona   188
irrelevance (waffle)   107
insanity (Franklin)   200
instant messaging   27, 59
when to   120
see also messages

intelligence and brevity 99, 124
interests versus belief 76
internal waffle 105
interruption by technology 37
intersecting monologues 112, 138
invention as human need 13
iPhone (milestone) 15
iPod (milestone) 13–14
issues and seriousness 197, 205

James, Oliver 24
James, William 84
jargon (waffle) 107
   alert list 211–13
Jones, keeping up with 28

Kane, Pat 17
Kingdon, Matt 160

language
   body 59, 120
   'machismo' 163
   understanding 103, 125
   and waffle 105
Larson, Doug 112
laziness and efficiency 168, 178
'less is more' 10, 100
Letterman, David 13
life
   separating from work 17
   work/life balance 18
life skills, missing out on 8
Lileks, James 32
listening 112–13
   developing skills 124
   and wisdom 112
locus of control 88
Lomasney, Martin 112

machines
   love of 31–3
   and nice/nasty things 192–3, 203
   see also technology
machismo language 163
Marketing Characters (Fromm) 24
marriage *see* relationships
mate/mum/grandmother test 119, 125
materialism and self-doubt 25
medium (method) and message 45, 72–3, 119–20, 129, 130, 138–9, 153
Meet (New Etiquette) 150
meeting up, when to 120
meetings
   and hand-held devices 59–60
   'topless' 60
messages
   fate of those sent 138
   pitfalls of 139
   see also email; instant messaging; medium; texting
Mies Van Der Rohe, Ludwig 82, 100
Mill, John Stuart 76
mobile phones/devices
   and driving 47
   inability to live without 39
   prevalence of 13
   social implications of 38
   and stress 53
movement and progress 168–9
multiple channels 56
multitasking 54
   and Rapid Sequential Tasking 171–2, 182

netbooks 15
New Communication
    Etiquette 150–1, 154
'next big thing' 82
Nintendo Wii (milestone) 14
'no', saying politely 115–17, 125
non sequitur 106

Obsessive Mobile Disorder 38, 95
online dating 31
outcome and output 169
out-thinking self 86
over-consumption 20–2
overnight test (writing) 145
overwork 21, 179
oxytocin 60

Paradox of Choice (Schwartz) 28
paranoia and products 27
Parris, Matthew 10
partnerships *see* relationships
Pascal, Blaise 99
pen and paper 44, 94
perfection and progress 165–6,
    181–2
persona and inner self 188
personal capitalism 24
Peters, Tom 57, 195
phobology 27
play 17–18
Play Ethic (Kane) 17
prejudices and thinking 84
presenteeism 21
product proliferation 27–8
progress
    and movement 168–9
    and perfection 165–6, 181–2
proof reading 145
purchasing and dopamine 20

quality of life estimates 22
Qualman, Erik 27, 77
quantitative versus qualitative
    tasks 159, 181

Rapid Sequential Tasking
    171–2, 182
relationships
    and social media 57
    and technology 31–2, 37–8
    and texting 42
repetition (waffle) 106
research for writing 141–2
resolutions
    being 74, 185–205
    communicating 72–3, 127–55,
        200
    doing 73, 157–83, 200–1
    talking 71–2, 97–125, 200
    thinking 69–95, 200
responsibility
    abdication of 90
    and locus of control 88
    and thinking 76
revolution
    need for personal 8, 65
    steps for 162–3
    to be able to cope 54
    see also resolutions; Taming
        Technology Tips
Revolution Test 62–5
ring tones and stress 53
road accidents and texting 42
Rotter, Julian 88
Russell, Bertrand 84, 90

Schwartz, Barry 27–8
self and seriousness 197, 205
self-definition 187–8

self-determination   74, 188
self-doubt and materialism   25
self-esteem
  and hand-held devices   57, 132
  reduction of   27
  and technology   29
self-worth, recapturing   24
Semler, Ricardo   177
Seven-Day Weekend (Semler)   177
Shaw, George Bernard   90
Shirky, Clay   10
Sky+ (milestone)   14
sleep, benefits of   178
Slingbox (milestone)   14
SMS (Short Message Service)
    41, 42
Snow, C.P.   27
social behaviour and texting   41–2
social interaction online   11
Social Learning Theory (Rotter)   88
social media/networks
  and asking 'why'   77
  distraction by   47, 163
  and relationships   57
  use by companies   59–60
social phobia and technology   27
social skills   56, 88, 132
Socialnomics (Qualman)   27, 77
speed of acquisition   25
spontaneous word dumps
    109–10, 124
spurious sources (waffle)   106–7
Sticky Wisdom (Kingdon)   160
straight thinking/talking   72, 99,
    102–3, 107, 119, 123, 195
stress
  and Ediction   8
  Freizeitstresse   50–1
  and ringtones   53

structure in writing   142–3
style, defining own   74, 79, 187,
    188, 201, 203
subject, researching   142
summary versus conclusion   144
survival strategy   20

'take on less, do more'   169, 182
talking   71–2, 200
  choice of medium   119–20
  face to face   56, 132, 138
  as resolution, 97–125
  straight   72, 99, 102–3, 107,
    119, 123, 195
Taming Technology Tips
  for being   203–5
  for communicating   153–5
  for doing   181–3
  for talking   123–5
  for thinking   93–5
tasks, managing   159–60, 163
techno rage   53–4, 154
technology
  addictive nature of   18
  and anonymity   56
  balancing with life   8
  as consumption   18
  and cowardice   133
  C.P. Snow on   27
  as curse   8
  as default option   155
  dependence on   29
  as devil   162–3, 181
  exponential spread of   11
  hiding behind   56–7, 132–3,
    153
  interruption by   37
  man's relationship with   7
  milestones of   13–15

morals about usage 147–8
pervasiveness of 50
and relationships 31–2, 37–8
right attitude to 192–3, 204
self defined by 187–8
and self-esteem 29
and social phobia 27
and thinking 77, 79–80
see also machines; Taming Technology Tips
television watching 11
test
   for Ediction 62–5
   overnight (writing) 145
   of understanding 119
texting 41–2
   when to 120
   see also messages
thinking 71, 200
   Barbara Castle on 90–1
   Bertrand Russell on 84
   conditions for 79–80
   and prejudices 84
   as resolution 69–95
   and responsibility 76
   straight 102–3, 123
   as turn-on 90
three Hs 195–7
Tipping Point (Gladwell) 82
title, choice of 142
Tom Tom Go (milestone) 14
Trout, Jack 59
Turnbull, Tony 17
turn-on, thinking as 90
Tyranny of Email (Freeman) 129

'understanding is everything' 10
understanding test 119
'unreasonable man' 90

up-gadgeting 20, 22
USB stick (milestone) 13

vacation and being 199–200, 204
vagueness (waffle) 107
variation and being 199–200, 204
video conferencing 120

waffle
   avoiding 123
   internal/external 105
   spotting 105–7, 124
wafflee/waffler 72, 105
   and conversations 112–13
wages *see* earnings
Watson, Jim 88
weak points (waffle) 107
web-browsing and thinking 53
webinars 120
wellbeing, recapturing 24
Whatever You Think, Think The Opposite (Arden) 166
Wilde, Oscar 178
WILFing 20
wisdom and listening 112
work
   overwork trend 21, 179
   separating from life 17
   versus play 18
Workaholics Anonymous 21
work/life balance 18
Write (New Etiquette) 150
writer's block 144–5
writing
   asking what/why 77, 141
   and Brachylogy 99
   concisely/clearly 141–5, 154
Wurman, Richard Saul 50